THE
BRAIN

A USER'S MANUAL

T0186943

THE
BRAIN

A USER'S MANUAL

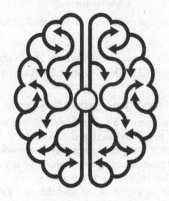

A SIMPLE GUIDE TO THE WORLD'S MOST COMPLEX MACHINE

MARCO MAGRINI

With an Afterword by Tomaso Poggio
Translated by Katherine Gregor

First published in Italian by Giunti Editore S.p.A., Firenze-Milano

First published in English in 2019 by Short Books
an imprint of Octopus Publishing Group Ltd
Carmelite House, 50 Victoria Embankment
London, EC4Y 0DZ

www.octopusbooks.co.uk
www.shortbooks.co.uk

An Hachette UK Company
www.hachette.co.uk

This paperback edition published in 2021

10 9

Cervello, Manuale dell'utente by Marco Magrini
Copyright © 2017 Giunti Editore S.p.A., Firenze-Milano
www.giunti.it

English translation copyright © Katherine Gregor

A CIP catalogue record for this bookis available
from the British Library.

ISBN: 978-1-78072-505-5

Cover design by Richard Green, richardgreendesign.co.uk

Image credits:
123RF: ©designua pp14, 19, 29
AdobeStock: ©Ben Schonewille p43
Shutterstock: ©vasabii, pp37, 52

Printed and bound in Great Britain by Clays Ltd, Elcograf S.p.A.

This FSC® label means that materials used for the
product have been responsibly sourced

MIX
Paper | Supporting
responsible forestry
FSC
www.fsc.org   FSC® C104740

To Jaja and Lilli

CONTENTS

GETTING STARTED

CONGRATULATIONS ON ACQUIRING THIS EXCLUSIVE PRODUCT, tailor-made for you. Please read this instruction manual carefully and keep it handy in case you need to refer to it later.

Your brain provides you with an extraordinary and unrivalled service. The availability of a sensory system for perceiving your environment and a nervous system that controls your motor apparatus, as well as an integrated consciousness that allows you to discern and decide, will sustain you for many years of existence.

As the famous inventor Thomas Edison said, 'The chief function of the body is to carry the brain around.' Just another way of saying that *we are* our brains.

The world is littered with millions of manuals. There are over 700,000 on the manualsonline.com website, one for any appliance you might have: from a deep-fryer to a lawn-mower, an electric toothbrush to a garage door. And yet, in this microcosm of informational odds and ends, there is no mention of the most important device each of us possesses.

The brain is a machine, in the sense that it performs a complex series of computations at the same time as decrypting in real time information coming from the numerous connected 'peripheral' senses, the most complex of which is our vision. The brain's response can be likened to an algorithm, as though the mind were the software that runs on the hardware of the encephalon (the material within your skull).

The brain is not, of course, a machine in the literal sense. It is neither hardware nor software. Some call it *wetware*, the wetness element highlighting the biological nature of the brain machine.

It is the most wonderful – and mysterious – fruit of evolution.

Wonderful because there is nothing in the entire universe that matches its complexity. It is made up of the very elements of the periodic table that form the stars, patiently arranged in such a way as to produce thought, speech and action, not to mention history, philosophy, music and science.

Mysterious precisely because science – an invention of the brain itself – is aware of its woefully inadequate grasp on its inventor. In fact, it is aware that it knows next to nothing.

Not only do we not know exactly how the brain works but a consensus hasn't even been reached on what it actually is. The brain's most amazing characteristic, consciousness (on which there is still no consensus, either), is a cerebral property that has sparked off centuries of furious debates – and not just among theologians and philosophers. There is no unanimity among scientists, for example, on the frequent loss of consciousness we call sleep: there are over 20 different theories to date on why the brain needs to fall asleep (while still carrying on working). As a matter of fact, we don't even have a consensus on the nature of sleep disorders or some of their unpleasant consequences, such as depression. We could go on listing our ignorance ad infinitum. And yet there is a lot we *do* know.

The early philosophers wondered whether the mind resided in the brain or in the heart, and influential thinkers like Aristotle leaned towards the latter. Nowadays, we know that the brain is the control centre of the nervous system in all vertebrates and a large proportion of invertebrates. We know about the stages of the brain's evolution. We know what it's made of. We know the genetic code of every cell and we can interpret that code. We have new technologies, like fMRI (functional magnetic resonance imaging) and MEG (magnetoencephalography), which allow us to observe cognitive activities as they are happening. We're advancing at breakneck speed in our understanding of the entire system.

A fridge instruction manual is compiled by the fridge manufacturer. Yet when it comes to the brain, the result of millions of years of evolution, it is only by drawing together clues from generations of earlier brains that we will be able, in the end, to solve the mystery and compile a perfectly comprehensive instruction manual. This is intelligence trying to understand itself, almost as though this were the inevitable evolution of evolution.

An exhaustive manual of everything we know about the brain, or think we know, would be monumental, and accessible only to a neuroscientist. This particular manual, however, is for the average human brain user. It's a collection of basic instructions about the most complex thing there is, and so will – we hope – be ideal for practical, everyday brain usage.

There is a famous quote – so famous, in fact, that it's attributed to at least three different authors: 'If the human brain were so simple that we could understand it, we would be so simple that we couldn't.'

Yet, in the end, of this we're certain: humankind will succeed. It's only a matter of time. Not tomorrow but in 20, 100 or 200 years' time, the brains of *Homines sapientes* will succeed in understanding themselves. But it will have taken a few hundred centuries of evolution for them to do so.

Technological progress, along with the extraordinary fruits of neuroscience over the last 20 years, confirms with every passing day the insight of Santiago Ramón y Cajal, one of the fathers of neuroscience back in the 1890s: 'Every man, if he so desires becomes the sculptor of his own brain.'

It's right and good that your brain, like that of every other user, should know the whys and wherefores of this self-sculpting process.

1.0 OVERVIEW

EVERY SINGLE SECOND OF THE DAY, including *this one*, your central nervous system is a laboratory housing millions of chemical reactions of which you're unaware. They are the language used by the brain to receive, process and transmit information.

The brain has long been seen as a machine. Since every idea is a child of its time, René Descartes compared it to a hydraulic pump, Sigmund Freud to a steam engine, and Alan Turing to a computer. Turing came closest – the brain isn't exactly a computer but there's an undeniable analogy between the two.

Both receive and transmit information by means of electrical messages. It's true that the messages are digital in the case of computers (expressed in the binary mathematics of zeros and ones) and analogue in the case of the brain (expressed in a variable arc of millivolts). But the brain's messages can also be binary, yes or no, on or off: if the sum of analogue messages it receives goes beyond a certain level, a neuron 'fires' and transmits an electrical impulse to connected neurons. If this level is not exceeded, nothing happens [Synapse, p18].

Both compute. But where most computers have a serial structure, that is, they compute following a preordained sequence, the brain operates using a parallel method, performing a huge number of computations simultaneously [Senses, p96].

Both require energy: the computer in the form of electricity, the brain in the form of oxygen and glucose [Nutrition, p84].

Both have an expandable memory: in the case of the first you must add or replace silicon memory banks, while the second needs only to multiply synaptic connections through study, exercise and repetition [Memory, p64].

Both have evolved over time: the computer at an exponential rate, doubling its processing power every two years, while the *Homo sapiens* brain – which originates from the brain of primitive invertebrates – has taken 500 million years to do so and has changed little over the past 50,000. It is, in fact, the same basic model that you, dear user, have been issued with [Topography, p35].

The brain's hardware, made up of intricately arranged atoms and molecules, contains approximately 86 billion neurons and weighs in at one and a half kilos. Since every neuron can fire and inundate thousands of adjacent neurons with thousands of signals as many as 200 times per second, it is estimated that the brain can perform up to 38 million billion operations per second. While it's nonsense that human beings use only 10% of their brains [Debunking Myths, p206], it is still pretty extraordinary that the brain manages all its activity while consuming less than 13 watts of energy per hour. No computer in the world has yet been able to beat the processing power of a human brain, let alone its extraordinary energy efficiency. And this is only the beginning.

The human body is constantly replacing old cells with new ones. All except the neurons, which are with you from the first to the last day of your life [End of Life, p211]. In the final analysis, they are the cells that make you what you are. Personality, skills and talents, erudition and vocabulary, tastes and tendencies, even memories of the past are written in your personalised neuron structure [Personality, p149]. So personalised that there is no brain anywhere in the world identical to yours, not even if you have a twin.

The brain is even able, within limits, to correct faults in its own hardware. When a particular area is damaged, the whole machine is often able to reprogram itself, moving the missing connections elsewhere and thereby restoring service [Strategies for the Brain, p224]. This is constantly happening on a small scale because, with the process of ageing, many neurons die and are never replaced. It also sometimes occurs on a large scale (as in the case of loss of vision, when the areas of the brain previously used for seeing put themselves at the disposal of other senses) [Vision, p102]. Don't tell this to a silicon processor, though, because even just one defective transistor in it can bring the entire device to a halt.

The brain reorganises itself via synapses, the estimated 150,000 billion connections between its neurons. In the words of the Canadian scientist Donald Hebb (in 1949): 'Neurons that fire together wire together.' This means that neurons that fire together connect and reinforce their reciprocal links. By creating new synapses, reinforcing old ones and cutting away those that are no longer needed, the brain constantly reorganises the relations between its neurons [Start-Up, p79]. A large number of brain functions, starting with learning, depend on this constant modification of synaptic connections and their strength. For centuries, it was assumed that the human brain – except for the childhood stage, when we learn to walk and talk – was essentially static and unchanging. It's only since the 1970s that we've been discovering that the exact opposite is true:

- In some cases the brain can self-repair.
- A child who is struggling in their studies can learn to learn. All they need is to be taught how to do so, and to be encouraged instead of humiliated [Learning, p162].

- Any bad habit, however unpleasant, can be kicked. Even a severe addiction, like gambling, can be controlled and subdued [Habits and Addictions, p193].
- An old lady can retain the memory power of a young adult as long as she doesn't stop learning and making a mental effort [Lifelong Learning, p216].
- By way of contrast, a condition such as post-traumatic stress disorder causes undesirable and enduring changes to the brain [Chronic Stress, p196].

The consequences of this ability to change, also called neuroplasticity [Plasticity, p69], surpass our imagination: the brain is a **powerful parallel computer** (capable of performing multiple operations simultaneously) that can adapt its hardware on its own.

The user of a working brain may discover that, through willpower – an act of will – they are able to adjust and fine-tune their own synaptic configuration [Control Panel, p153], which forms the basis of their own life. Until we meet an alien of superior intelligence, the brain of *Homo sapiens* remains the most complex, amazing and brilliant thing in the universe. Its very complexity enables its neurons to produce thought, intelligence and memory – all tailor-made to the user. It is amazing that this biological machine is still widely superior, in power and efficiency, to all the machines in the world – and it's thrilling to take a ride on it.

Warning: in some cases, a faulty working of the brain machine reflects an underlying medical condition that is beyond the merely informative remit of this manual [Disclaimer, p250] and requires professional advice and treatment [Malfunction, p201].

1.1 TECHNICAL SPECIFICATIONS

Weight (average)	1350	grams
Weight in comparison to total body weight	2	per cent
Volume (average)	1700	millilitres
Length (average)	167	millimetres
Width (average)	140	millimetres
Height (average)	93	millimetres
Average number of neurons	86	billion
Neuron diameter	4-100	microns
Electrical potential of resting neurons	-70	millivolts
Sodium pumps per neuron	1	million
Number of synapses	> 150,000	billion
Grey matter / white matter ratio in cortex	1:1.3	
Neurons / neuroglia rate	1:1	
Number of neurons in cerebral cortex (women)	19.3	billion
Number of neurons in cerebral cortex (men)	22.8	billion
Loss of neurons in cortex	85,000	per day
Total length of myelinated fibres	150,000	kilometres
Total surface of cerebral cortex	2,500	square centimetres
Number of neurons in cerebral cortex	10	billion
Number of synapses in cerebral cortex	60,000	billion
Layers of the cerebral cortex	6	
Thickness of the cerebral cortex	1.5-4.5	millimetres
Volume of cerebrospinal fluid	120-160	millilitres
pH of cerebrospinal fluid	7.33	
Number of cranial nerves	12	
Blood flow	750	millilitres/second
Oxygen consumption	3.3	millilitres/minute
Energy consumption	> 12.6	watts
Maximum speed of electrical impulses	720	kilometres/hour
Function temperature	36-38	degrees Celsius

1.2 SYSTEM VERSION

Your brain is Version 4.3.7 (G-3125)* of a nervous system that has undergone careful genetic refinement spanning hundreds of millions of years of evolution, designed to provide you with the full experience of human life on this planet.

- For upgrades (currently not available), please see the section on Future Versions [p233].

---

*   **Version 4.3.7 (G-3125)** is composed as follows:
    4: invertebrate / vertebrate / mammal / primate
    3: hominid / Australopithecus / *Homo*
    7: *Homo habilis* / *Homo ergaster* / *Homo erectus* / *Homo antecessor* / *Homo heidelbergensis* / *Homo sapiens* / *Homo sapiens sapiens*
    G–3125: number of generations (estimate) from the advent of modern man's brain (*Homo sapiens sapiens*) to your brain.

2.0 COMPONENTS

YOUR BRAIN LOOKS LIKE A SINGLE object, but it isn't. It is often described as a network of neurons but that's not right either. If anything, we could say it's a network of networks of networks.

Every single brain cell, or **neuron** [p12], could be viewed as a basic microscopic network. Each cell is ruled by the genetic instructions it contains and operated by millions of ion channels, sodium-potassium pumps and other chemical devices. But this single unit, however sophisticated, is useless on its own. It has power only in conjunction with other neurons. It's no coincidence that the processing power manifests not so much in individual brain cells as in the connections between them, or **synapses** [p18].

Neighbouring neurons are organised together into operational units known as **nuclei**. There are over fifteen nuclei contained in the hypothalamus alone [p46], which is only the size of an almond. Other nuclei link up in a chain in order to form **brain circuits** that control specific functions, such as sleep or attention. Just as many neurons form a circuit, many circuits join forces to create diverse features, such as language and empathy. It is this monumental network of networks that generates consciousness [p126] and intelligence [p72].

This nuclei system wouldn't work without a parallel network: the **glial cells** [p28]. The glial cells feed, oxygenate and cleanse the neurons. Even more importantly, they regulate the extraordinary speed of the **axons** – the long fibres extending from neurons [p17] – by coating them in a white fat called myelin, which, in a nutshell, amplifies their signal [p32]. The **cerebral cortex**, which, unlike nuclei, is organised into six hierarchical layers, owes its effectiveness to the ability to transmit high-speed signals over long distances. The total length of the myelin-coated fibres in your brain is estimated to be about 150,000 kilometres: almost four times the Earth's circumference at the Equator.

In this monstrously complex network, the **right** and **left hemispheres** of the brain (which regulate the opposite sides of the body) work together, as do the four lobes and the various operating areas of the cortex, the seat of many of the functions – thought, language and consciousness among them – that make us human. Each component has its own network, its own place in the hierarchy, and its own particular mission. Each is an essential part of the amazingly complex super-network that is the human mind – a mind that has dreamed up the Great Pyramid of Giza, da Vinci's *Mona Lisa* and Mozart's *Requiem*, and has figured out gravity and natural selection.

2.1 THE NEURON

An average human male is made up of approximately 37,000 billion cells – an exorbitant number of biological bricks are required to build a human specimen like you. In this intricate construction of bone, blood, liver and skin cells, one group takes centre stage: the neurons.

It's estimated that your brain contains 86 billion of these astounding brain bricks, accompanying you from birth to death. Forming an intricate network composed of hundreds of thousands of billions of connections, they transmit electrical and chemical impulses to one another at the rate of hundreds of kilometres per hour within a time frame of a few milliseconds.

It's this network that, at this moment, allows you to read and understand. It's this network that enables you to create memories, ideas and feelings – and much, much more.

The central body of each neuron, called the **soma**, is of infinitesimal dimensions (the smallest is 4 microns wide, 4 millionths of a metre) and yet in some cases the cell of which it forms part can stretch to several centimetres due to its axon. Every neuron has one axon, which acts a little like a transmission cable, carrying information out of the cell to other neurons. On the inputing side, there are other, shorter extensions: the **dendrites**. Each neuron has multiple, many-branched dendrites, which, like receiving cables, carry information into the cell.

Neurons can take on many forms – there are over 200 kinds. The differences between them lie in the functions they perform within the brain network. **Sensory neurons** (also known as afferent neurons) channel incoming signals from organs like the eyes and skin towards the central nervous system. **Motor neurons** (also known as efferent neurons) carry motor-type signals from the central nervous system to peripheral organs, all the way down to your toes. The remaining type – **interneurons** – create the wonder of intelligence through a network of astonishingly intricate connections.

The communication between neurons is effected by a series of chemicals, called **neurotransmitters** [p20], that start up on orders from the cell. The order arrives via **action potentials**, variations of just a few milliseconds in the electric tension travelling through the cell, that trigger the release of neurotransmitters (such as dopamine, serotonin and noradrenaline) at the synapse, the junction between cells. The neurotransmitter travels across an infinitesimal space between the adjacent cells, the **synaptic cleft**. When a neuron emits an action potential, it 'fires', sending a message to receptor neurons, inciting them to fire in turn or inhibiting them to silence.

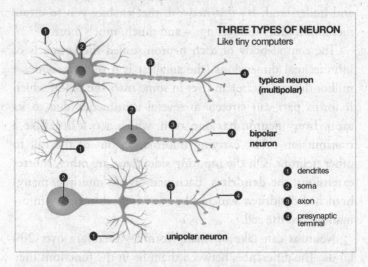

THREE TYPES OF NEURON
Like tiny computers

typical neuron
(multipolar)

bipolar
neuron

unipolar neuron

1 dendrites
2 soma
3 axon
4 presynaptic
terminal

There is a system that runs in parallel to this communication system, one which involves electrical, rather than chemical, synapses. It's much faster, digital (the signal is only on or off), doesn't have long-distance axons and involves only neighbouring neurons. It is based around the nuclei, or groups of neurons, organised in specialised **neural pathways** – rather like many orchestras playing different scores. The neurons are connected along these pathways by chemical synapses, but they are also linked by electrical synapses that coordinate the activities of the orchestra, made up of millions of neuron musicians. The electrical impulses between these cells, which are constant and synchronised, are known as brainwaves, or neural oscillations. Brainwaves are sent out regularly at various frequencies (measured in hertz, or cycles per second) and involve different areas of the brain, depending on the degree of wakefulness – from deep sleep to excitement.

It is now clear that brainwaves, studied initially because of their role in governing sleep [p87], play a key role in neuro-transmission and cognitive and behavioural functions. At the very least, this is because they synchronise and beat time for each neuronal orchestra. But perhaps they do even more. It's thought that brainwave rhythm could even be linked to the mystery of consciousness [p126], though there's no decisive evidence for this.

WAVES	HERTZ	ASSOCIATED WITH	EXAMPLE
Delta	1-4	deep sleep (non-REM)	unconscious state, body immobilised
Theta	4-7	REM sleep, meditation	sleeping and dreaming of a nice holiday away
Alpha	7-12	calm, relaxation	thinking that, actually, it's time to take a nice holiday
Beta	12-30	concentration, mental effort	organising two weeks of planes, hotels and car hire
Gamma	30-100	heightened attention, anxiety	discovering that your current bank account is overdrawn

2.1.1 The Dendrites

Imagine a forest thicker and more intricate than any you've come across. Billions of trees with hundreds of billions of branches and thousands of billions of leaves, all connected to one another in a way that allows communication from one side of the wood to the other. An enchanted forest. Partly because it's extraordinarily beautiful and partly because of the magic it performs. These are the neuron's dendrites – so reminiscent of trees that they are named after them (*dendron* means 'tree' in Greek). The dendrites, the receptor terminals of the nerve cells, stretch into an explosion of branches and foliage, which, depending on the type of neuron, can resemble a pine or an oak, a baobab or a redwood.

Some dendrites even have leaves, or spines. In the same way that the leaves of a tree are the receptor terminals for sunlight, triggering photosynthesis, dendrites and their spines are the receptor terminals for information arriving from the transmitter terminals of other neurons.

As in any forest, the neuron's branches aren't stationary. In the last ten years it has become clear that dendrites and their spines play a key role in neuroplasticity – the brain's ability to constantly rearrange its neuronal connections [p69]. Neuroplasticity is not an abstract quality: the brain physically changes with the growth of new branches and new leaves, or the loss of those that have dried up. The growth and adaptation of new spines and new dendrites, and the strength or weakness of individual synaptic connections, govern learning and memory [pp162, 64].

2.1.2 The Soma

The neuron's management office, called the soma, is the cell's central body. The dendrites and the axon branch off from it. It generates the energy needed by the cell, manufactures its parts and assembles them. It is contained by a membrane made up of fats and chains of amino acids that protects the neuron from the external environment. Inside are a host of specialised structures, starting with the nucleus, which acts both as an archive and a factory: it stores DNA, which contains all the information for building the proteins essential for survival, and manufactures RNA, the template required to synthesise them.

Structures called **mitochondria**, present in all cells, use oxygen and glucose to generate the fuel, known as ATP (adenosine triphosphate), but in gargantuan amounts: no cell has a larger appetite than the neuron.

2.1.3 The Axon

A neuron has many receptor dendrites, but it has only one axon – just one pathway for transmitting a signal to its peers.

Whereas dendrites reside in the area immediately surrounding the soma, within a radius of just a few microns, the axon can stretch as far as tens of centimetres which, at this scale, is an impressive distance.

Whereas dendrites tend to branch like trees, the axon keeps a constant diameter until, near its tip, it splits into many small transmitter branches, called **axon terminals**, via which it makes connections with other neurons.

There is another significant difference: while the chemical signal that reaches dendrites can be intense, weak, or anything in between, the electrical signal that travels through the axon is either there or not, on or off. In this respect, we could say that dendrites are analogue devices while the axon is essentially digital.

The mission of the axon isn't just to send information long distance; it's also to send it high speed: in extreme cases, this can reach 720 kilometres per hour, or 200 metres per second. The speed depends on the diameter of the axon and, above all, on the thickness of the **myelin sheath** that insulates it from outside interference. There's a direct relationship between the amount of myelin and the extent to which the axon is used [p162]. Unlike motorways, which get worn out by traffic, neural pathways grow stronger with the circulation of electrical impulses, acquiring thicker myelin.

The transmission begins in the **axon hillock**, the point

where the soma of the cell narrows to form the axon. It's more or less the centre of the entire computational process – where additions and subtractions take place. If the result of the computation goes over a certain electrical threshold [p21], it prompts the neuron to fire, and shoot an action potential.

The myelin sheath around the axon has tiny, regular gaps (called **nodes of Ranvier**) where the axon is exposed. In these nodes, a system of canals allows sodium ions to enter and exit the cell, amplifying the action potential by enabling it to leap from one node to the next at a speed that would not otherwise be possible.

Myelin is deeply involved in human intelligence [p162]. The loss of myelin in diseases such as multiple sclerosis, leads to impairment in the transmission of signals, and therefore the correct functioning of the brain machine.

While it's the large concentration of neuron cell bodies that creates the colour of the so-called **grey matter** in the cortex, **white matter** is given its colour by myelin. Axons, which make up the white matter of the area where the two cerebral hemispheres join (the **corpus callosum**), occupy more space than the somas, dendrites and spines put together.

2.1.4 The Synapse

After the dendrites, the soma and the axon, we reach the neuron's final stop: the synapse. This is where the axon terminals of one neuron (**presynaptic**) meet the branches, leaves or body of another neuron (**postsynaptic**). There is an infinitesimal gap (between a 20 and 40 billionth of a metre) between the two neurons: the **synaptic cleft**. It is there that the enchanted forest of neurons comes alive: the precise spot where brain cells speak to one another, using the language of chemistry.

The axon terminal stores neurotransmitters in small spheres called **vesicles**. At the command of an action potential, the neurotransmitters are released and travel across the synaptic cleft where they come into contact with the receptors of the second neuron, thereby causing a signal to be triggered. This is only one link in a marvellous chain of signals travelling through your brain millions of times a day, allowing you to think about the past, plan your future and move your legs in the present.

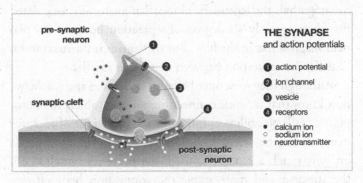

pre-synaptic neuron

synaptic cleft

post-synaptic neuron

THE SYNAPSE
and action potentials

1 action potential
2 ion channel
3 vesicle
4 receptors
• calcium ion
○ sodium ion
• neurotransmitter

A neuron can be connected to tens of thousands of other neurons, even in remote areas of the brain. Pyramidal neurons, the most common cells of the cerebral cortex, have between 5,000 and 50,000 receptor or postsynaptic connections. Purkinje cells, another type of neuron, can have up to 100,000. According to some estimates, about 150,000 billion synapses lurk in a young adult brain. But, although it has been possible to estimate the average number of neurons in a human brain [p12], accurately calculating the number of synapses poses an insurmountable challenge.

The heart of the matter is not so much accurate quantification as the explosive power of the neural network, with its exponential mathematics.

Let's take a hypothetical neuron that synaptically speaks

with 'only' 1,000 other neurons. Each of these is potentially connected to another 1,000, so that during the second leg of its journey – within just a few milliseconds – the information reaches a million cells (1,000 x 1,000). The third time around, if they were connected to another 1,000, the total would be one billion (1,000 x 1,000 x 1,000). This calculation is simplistic because communication is much more complex between the different kinds of cells, nuclei and neural pathways. Still, it gives us an inkling of how powerful the machine is. János Szentágothai, the legendary Hungarian anatomist, calculated that there are only 'six degrees of separation' between any two neurons, just like in the film. But six degrees is the maximum. Usually, the separation between neurons is smaller.

Although they were once believed to be fixed and stable, we now know that synaptic connections are also subject to neuroplasticity. Their ability to influence the behaviour of receptor neurons can be strengthened or weakened. It all depends on how much a synapse is used: the more it's switched on, the stronger and more stable the connection between two brain cells becomes [p5]. This phenomenon, called **long-term potentiation (LTP)**, has important practical implications for learning systems [p162] and memory [p64] – and is also central to the habituation and addiction process [p193].

2.2 NEUROTRANSMITTERS

The brain speaks the language of neurotransmitters. At any given moment, whether you're reading a book or admiring a view, there's a chemical tempest raging in your encephalon. Millions of microscopic molecules are leaving the vesicles of the neurons and travelling across the synaptic cleft, each carrying their own chemical message. The brain uses these neurotransmitters to instruct the heart to beat, the lungs to breathe, the stomach to digest. These molecules are also needed to order your brain to sleep or pay attention, learn or forget, get excited or relax. All this behaviour – from the most rational to the most unconscious – is mediated through an army of neurotransmitters and their complex interactions. Over a hundred types have been identified so far but that doesn't mean more won't be discovered.

Synaptic messages can be either excitatory or inhibitory to a variable degree, depending on both the neurotransmitters that set off from a neuron and the receptors that capture them in the adjacent neuron. Each receiving neuron can be connected to thousands of other neurons through as many synapses, and therefore can simultaneously receive thousands of impulses through their synapses. Excitatory and inhibitory messages are 'added up' inside the cell, which, thanks to a sophisticated pumping system that regulates the access and flow of sodium and potassium ions, maintains its membrane at a stable 'resting' electrical potential of -70 millivolts. Excitatory neurotransmitters contribute to making the voltage of the neighbouring membrane positive, while inhibitors tend to the negative. If the net result of the adding up goes over a certain voltage (usually -30 millivolts), the nerve cell fires and triggers the action potential, the electrical impulse that runs down the axon to command the release of another volley of neurotransmitters. If, on the other hand, it does not go over it, everything stops there.

The mathematics of neurotransmission reaches far beyond simple voltage calculation, because messenger molecules perform their tasks in combination with or in opposition to one another. The resulting range of possibilities is so vast that it includes reason, memory and emotion.

The Swedish researcher Hugo Lövheim suggested a model

for the interactions of the neurotransmitters serotonin, dopamine and noradrenaline. According to his classification, the levels of these three molecules determines your basic emotions. Anger, for instance, has high levels of dopamine and noradrenaline, and low levels of serotonin.

	Serotonin	Dopamine	Noradrenaline
Shame	▽	▽	▽
Suffering	▽	▽	▲
Fear	▽	▲	▽
Anger	▽	▲	▲
Disgust, hatred	▽	▲	▽
Surprise	▲	▽	▲
Wellbeing, pleasure	▲	▲	▽
Interest, excitement	▲	▲	▲

▲ = high ▽= low

Naturally, reality is much more complex than this model suggests – if nothing else because of the mutual interaction of a huge array of other messenger molecules. Also because of a far from negligible detail: the magazines of the synaptic machine guns, the vesicles, are not always readily loaded with cartridges.

The availability of neurotransmitters is not unlimited. After they are tied to the postsynaptic receptor, they are promptly deactivated and recycled; or taken back into the vesicles, thereby reloading them (which is what we call **reuptake**); or else removed or even destroyed. Your brain could be the victim of a low supply of some types of molecule. Poor nutrition [p84], chronic stress [p196], medications, drugs, alcohol, and also genetic predisposition [p202] have an influence on the stock of neurotransmitters and can compromise the optimal functioning of the brain machine.

Some neurotransmitters, such as dopamine, serotonin, acetylcholine and noradrenaline also act as **neuromodulators**. If neurotransmission can be compared to a laser that strikes postsynaptic neurons with precision, then neuromodulation is like a spray. All it takes is for a few neurons to secrete neuromodulators for many other neurons to be affected across a wide area, thereby modulating their activity. Finally, **hormones** such as testosterone and cortisol can influence neurotransmission, further complicating all this busy synaptic activity.

GABA

The job of this neurotransmitter is to inhibit. *Gamma*-aminobutyric acid, better known as GABA, is the main inhibitory component of a synapse. Scarcity of this chemical can lead to anxiety. When it is abundant, it relaxes and facilitates concentration. It's no coincidence that medications that increase the availability of GABA – such as Valium – have a relaxing, anti-convulsive and anti-anxiety effect.

Glutamate

This is the excitatory neurotransmitter par excellence, and the most common. In large quantities it's highly toxic for neurons. It's essential to cognitive processes such as memory and learning, and also contributes to normal brain development.

Adrenaline

Otherwise known as epinephrine, this is the 'fight or flight' neurohormone produced in situations of stress. It increases the blood flow to muscles and the flow of oxygen to the lungs, preparing you to do battle or to flee. This hormone is produced by the adrenal glands.

Noradrenaline

Also known as norepinephrine, this is an excitatory neurotransmitter. It regulates attention and the 'fight or flight' response, also increasing the heart rate and consequently the blood flow to muscles. In large amounts, it causes anxiety, while low levels are associated with poor concentration and sleep problems.

Serotonin

This neurotransmitter contributes to feelings of wellbeing and – in its capacity as an inhibitor – balances any possibly excessive excitatory neuron activity. It regulates pain, digestion and, together with melatonin, sleep mechanisms. Low serotonin levels are associated with depression and anxiety, so much so that many antidepressants work by increasing its levels [p202]. Serotonin is also naturally produced through physical exercise and exposure to sunlight.

Dopamine

This is the neurotransmitter superstar. The reason it enjoys such good press is probably that it's the molecule connected to the reward and pleasure perception system [p141]. Excitatory but with inhibitory potential, it's involved with habituation and addiction mechanisms [p193], yet it would be inaccurate to think of it only as the 'pleasure molecule'. In light of recent discoveries, we could say that it is the neurotransmitter of the will. It is also indispensable to strategic functions, such as attention and movement. The distribution of neurons equipped with dopamine and related brain circuit receptors has helped identify a **dopaminergic system** with eight 'pathways' that distribute the molecule, also with a neuromodulating effect. The three most important pathways – mesolimbic, mesocortical and nigrostriatal – all originate from the **mesencephalon** and lead to the upper levels of the brain.

Acetylcholine

This is the most abundant neurotransmitter in the human body. In the peripheral nervous system, it stimulates muscle movement, and in the central nervous system, it contributes to excitement and reward, as well as performing important functions in the learning process and in neuroplasticity. Also a neuromodulator, acetylcholine is present in cerebrospinal fluid [p33] and consequently has an effect on disparate neuron regions.

Oxytocin

All you need to do to increase the amount of oxytocin is kiss, hug or have sex. Alternatively, you can breastfeed, and this hormone (which is also a neurotransmitter) will irrigate the mother's brain as well as the child's. In other words, it takes two to produce oxytocin naturally. Called the attachment molecule, because it promotes a sensation of wellbeing that encourages the creation of emotional connections, it is also believed to play a role in a range of physiological functions, from erection to pregnancy, from uterine contractions to milk production, from social bonds to stress. The presence or lack of oxytocin affects how helpful you are to others, as well as your psychological stability. Synthetic oxytocin, available for sale in some countries in inhalant gas form, is used as a recreational drug.

Vasopressin

A neurotransmitter and neuro-modulator, vasopressin is made up of nine amino acids. Besides performing more prosaic tasks, such as acting as a diuretic and constricting blood vessels, this molecule has a strategic role in the

human brain: the continuation of the species. Vasopressin intervenes in social behaviour mechanisms, sexual drive and couple attachment. The example of the *Microtus ochrogaster*, a distinctly monogamous vole (a rarity among mammals) that resides in the US Midwest, is famous: when lacking vaso-pressin, even this rodent ends up getting a divorce.

Testosterone, estradiol and progesterone

Just as the central nervous system uses neurotransmitters for sending its messages, the endocrine system uses hormones. So-called sex hormones, such as testosterone (male; picture above), estradiol and progesterone (female) play a key role in both the embryonic development of the brain and the small but significant differences between the adult brains of the two available models [p178] — men and women produce both testosterone and progesterone, but in totally different proportions.

Cortisol

Like the 'sex hormones', cortisol is not a neurotransmitter in the strict sense of the word — but it's nevertheless a mole-cule capable of influencing the brain machine significantly. Produced by the adrenal glands (by order of the hypothalamus [p46] as part of the complex response mechanism to protracted danger) [p116], cortisol is also called the 'stress hormone' [p196]. When cortisol levels remain high for a long time, the hippocampi [p45] suffer damage, and the brain ages at a faster rate. Cortisol also interferes with the learning process [p162].

Endorphins

We use the plural because endorphins are an entire category of opioids ('endogenous morphine' – that is, produced inside the body), which inhibit pain signals, ease them, and can give a sense of wellbeing, or even euphoria. They are released during physical exercise [p91] and sexual activity, and in response to pain. Certain foods, like chocolate, also trigger the release of endorphins.

2.3 GLIAL CELLS

Neurons, the intelligence cells, represent only one part of the brain mass. The rest is made up of another category of cells, called *glial cells* (from the Ancient Greek γλοία, 'glue'). First described at the end of the nineteenth century, for a long time they were thought to be a kind of scaffolding, or superglue, that supported the neurons. This view has radically changed since the 1980s, however, partly thanks to Albert Einstein. We can now guarantee that your brain is not filled with glue.

Although the greatest physicist of all time did not work in the field of neuroscience, he did leave an involuntary bequest after his death. While in the process of performing a post-mortem on Einstein's body in 1955, a doctor at Princeton Hospital, a certain Thomas Stoltz Harvey, came up with the idea of stealing the genius's brain. This theft – justified in the name of science – caused him a lot of problems.

Even so, for a long time there didn't seem to be anything

special about Einstein's brain. It was only 30 years later that Professor Marian Diamond of Berkeley University managed to unearth a peculiar trait in one of four different samples. In the parietal lobe region, where mathematical reasoning, spatial awareness and attention are transmitted, Einstein's glial cells seemed more numerous than normal. Although this discovery has since been disputed, it opened the floodgates to new research on glial cells.

We know now that glial cells are jacks of many trades. It's true that, as was once believed, they act as scaffolding: they surround the neurons and keep them in their place. As early as during the embryogenesis stage – when the brain starts assembling itself inside the placenta – the glial cells regulate the migration of neurons and produce the molecules that determine the branching of dendrites and axons. But they also constitute a pantry, feeding and oxygenating the neurons. They work as electricians, building the myelin sheath that regulates the transmission of action potentials through axons. In addition, they are most definitely rubbish collectors, keeping pathogens at bay and wolfing down neurons that are no longer active.

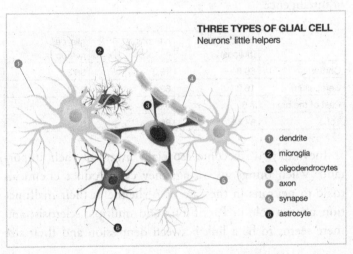

THREE TYPES OF GLIAL CELL
Neurons' little helpers

1. dendrite
2. microglia
3. oligodendrocytes
4. axon
5. synapse
6. astrocyte

What an extraordinary number of strings to their bow, without which the human brain would not work as it should. Recent studies also attribute to glial cells the ability to communicate among themselves chemically. Unlike neurons, they are capable of mitosis – that is, dividing and reproducing.

Many sources claim that there are five to ten times more glial cells than neurons. However, a recent study has debunked this myth, suggesting that it's more likely to be a ratio of 1:1. According to this complex calculation method (as usual, disputed by some), there are 86 billion neurons and 84.6 billion glial cells in the entire brain, though with significant differences in their ratio in the various brain regions. In the cerebral cortex, the part of the brain that most distinguishes *Homo sapiens* from other species, glial cells are almost four times as numerous as neurons. And in the white matter of the cortex, where the majority of myelinated axons are located, there are ten times more glial cells than neurons. Without even having to trouble Albert Einstein's poor brain, it is obvious that glial cells play a key role in the generation of intelligence.

	glial cels (billions)	neurons (billions)	glial cell / neuron ratio
Cortex	60.8	16.3	3.73
Cerebellum	16.0	69.0	0.23
Rest of the brain	7.8	0.8	9.75
Total	84.6	86.1	0.98

Inevitably, we become especially aware of their importance when things go wrong: they can produce chemicals toxic to neurons in the case of Alzheimer's; their malfunction plays a role in Parkinson's and multiple sclerosis; and there seems to be a link between depression and their size and density. In general, one could say that glial cells' primary task is to maintain homeostasis, the state of chemical–physical balance in an organism. In other words, to protect the status quo.

2.3.1 Microglia

The brain is essentially insulated from the rest of the world by the blood–brain barrier [p33], which prevents the entry of large-size infectious agents. If, however, a foreign body succeeds in penetrating it, microglia – spread throughout the brain and the spinal cord – go on the offensive to destroy the invaders and reduce the inflammation they have caused. Microglia – a type of glial cell – are small, but they eat a lot. Their task is to keep under constant control the surrounding environment and the health of neurons, other glial cells and blood vessels.

2.3.2 Astrocytes

We all know that the stars in the sky are just huge, round balls of gas. And yet in many cultures, they are drawn with five, six or seven points, because of the atmosphere's optical diffraction or – quite simply – the observer's astigmatism. It is because of their vague resemblance to pointed stars that the most common glial cells are called astrocytes. They're like a parallel universe, a galaxy of stars in the encephalic microcosm.

These vital star-shaped (in most cases) cells keep the brain together, contributing to the complex architecture of the cerebral mass. Not only that, they maintain homeostasis. They store and distribute energy. They defend the brain from external molecular attacks. They recycle neurotransmitters. They wrap around synapses, checking that transmission is working correctly. The list could go on.

2.3.3 Oligodendrocytes

Every fan of HiFi – as 'High-Fidelity' reproduced music was once called – knows that the cables that connect the record player to the amplifier, and the amplifier to the loudspeakers, must be well insulated in order to transmit frequencies faithfully and avoid interference. This is something that oligodendrocytes (from the Greek 'cells with few branches') also seem well aware of, since their job is precisely to insulate axons in such a way as to ensure that the electrical impulse transmission system works correctly.

This is no mean feat. Every oligodendrocyte can easily connect with 50 or so different neurons, coating axons with a sheath made of several layers of myelin – the mixture of fats and proteins that has altered the course of evolution. The fact that you can now enjoy a speed of neuron electrical impulse transmission of up to 200 metres per second is because the myelin sheath produced by oligodendrocytes allows axons to give a true HiFi performance.

2.4 OTHER COMPONENTS

Besides neurons, glial cells and the highly complicated molecular microcosm that allows them to function, your brain is equipped with two further devices that are essential for its survival. These relate to blood and water – and their respective plumbing systems.

2.4.1 The Blood–Brain Barrier

Long before human beings invented aquarium filters, air-conditioning filters or even cigarette filters, evolution had already equipped their brains with a far more ingenious filtering system, called the blood–brain barrier.

The cells that line the central nervous system have narrow junctions that allow only certain molecules through, to be carried into the brain by the bloodstream. The road is blocked to all undesirable molecules – toxins and bacteria in particular.

Thanks to the filter you have been equipped with, brain infections are a rare event. The trouble, though, is that in case of infection, the blood–brain barrier doesn't allow in pharmaceuticals made of large molecules or even the vast majority of those with small molecules. Research is trying to address this problem by developing nanomolecular drugs (in the region of one billionth of a metre) capable of penetrating the brain filter.

2.4.2 Cerebrospinal Fluid

The brain floats. A special fluid, transparent and colourless, mainly composed of water, acts as a liquid cushion to prevent it from being crushed by its own 1,350-gram weight. It has been calculated that, floating in this fluid, the brain has an effective mass of only 25 grams.

The cerebrospinal fluid (which, as the name suggests, is also found in the spine) has four other essential functions:

- It protects the brain in case of bumps – or at least tries to (no olden-days footballer would have headed all those leather balls if he'd known precisely what was inside his skull).

- As the principal component of the glymphatic system (so called because it resembles the lymphatic system but is regulated by glial cells [p28]), it does the housework. In a nutshell, this means washing away brain detritus – especially during sleep – through the opening of canals formed by the contraction of glial cells.

- Thanks to its regulating mechanisms, even though the brain produces about half a litre of cerebrospinal fluid every day, constant turnover ensures that only 120-160 millilitres are ever simultaneously circulating between the brain and the spine. Without this self-limiting feature, the pressure inside the skull would become so high that the brain would not have an adequate blood supply.

- Finally, the fluid – kept in the **meninges**, the membranes around the brain – is produced in the ventricular system, a complex of four interconnected brain cavities, and is eventually absorbed into the blood. By virtue of this circle of exchange, it also maintains the chemical stability of the floating brain machine.

3.0 TOPOGRAPHY

NOT EXACTLY 525 MILLION YEARS AGO, but thereabouts, the first vertebrates appeared on this planet. That is, life forms that have a spine that stretches down the entire body. Over the following millions of years, a process of encephalisation concentrated vertebrates' brain functions in the front region, the head.

The brain, which grew increasingly large, complex and efficient over the lengthy timeline of prehistory, is divided into two almost symmetrical hemispheres – connected at the centre of the **corpus callosum**, a sheaf of nerve fibres. Their function is nearly always asymmetrical: the areas dedicated to language, for instance, are usually situated in the left hemisphere in right-handed individuals (95% of the time), and a little less often (70%) in the left hemisphere in those who are left-handed.

The brain's topography – how exactly it is mapped out – has much to do with evolution: the human brain has inherited the structure of the brains that have preceded it.

The triune (three-in-one) brain theory, put forward in the 1960s by American scientist Paul MacLean, renders the ancestral origins of the world's most complicated machine easily understandable (although some aspects of his theory are now obsolete, or refined almost beyond recognition). In any case, it shows how evolution has found it preferable to add extensions and improvements to the structure of the brain rather than start each new version over from scratch. It goes as follows.

Deep in the cellar of your brain there is a part that is of reptilian origin. It is the oldest and smallest of the three parts, controlling the vital functions, from breathing to heartbeat, from body temperature to what we usually call instinct, including ancestral, territorial behaviours. These are essential brain operations that take place without the need for thought or will.

Then there's the brain that developed in particular with the advent of mammals, called the **limbic system**. It plays a key role in emotion, motivation, behaviour and long-term memory. These are cerebral activities that favour aspects of sociability typical of mammals, such as reciprocity and the ability to feel affection.

Finally, in the attic, we have the **cerebral cortex**: six layers of grey matter wrapped around the encephalon, which – especially developed in primates, apes and, most dramatically, in human beings – manage consciousness, thought, language, and all the things that make your brain stand out, such as predicting and programming future events [p60].

Naturally, all three brain regions have continued to evolve over millions of years, following the genealogical branches of the animal world. They are not sealed and separate from one another like a Russian doll, but tightly connected by an intricate network of neural pathways.

Of course, you do not have a reptilian brain – only a few cerebral structures that originate from a very distant ancestor you and reptiles have in common. There aren't literally three brains but just one.

3.1 THE 'REPTILIAN' BRAIN

Your brain's most distant ancestor dates back about 500 million years and first appeared under water as a primordial brain made up of a few hundred primitive neurons. As years – millions of years – went by, these primordial brains grew in complexity, as did the complexity of the underwater animals that contained them. When a few species emerged from the water to colonise the earth, they needed to contend with increasing complexity to survive – and a brain evolved about 250 million years ago which we could call a reptilian brain. Installed in increasingly sophisticated amphibian species, it became standard equipment in all the brains to come: its basic design can be found in modern reptiles as well as modern mammals, including *Homo sapiens*. With the huge differences accumulated over millions of years of evolution, naturally.

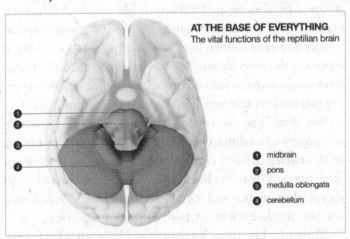

AT THE BASE OF EVERYTHING
The vital functions of the reptilian brain

1. midbrain
2. pons
3. medulla oblongata
4. cerebellum

The oldest and innermost part of your brain is composed of the **brainstem** and the **cerebellum**, and controls vital functions such as heartbeat, balance, breathing and body temperature.

It is a fully reliable component of the brain machine, able to work 24 hours a day on totally automatic mode, without the need of any intervention on the part of the user: nobody ever forgets to breathe.

3.1.1 The Brainstem

Seen from the base, the brain begins with a kind of groove – a 'nerve canal' we could say – that connects it to the rest of the body. This is the brainstem, which is so fundamental to cerebral processes and their evolutionary history that it regulates, among other things, the beating of the heart, sleep and hunger.

All the information that travels from the body to the brain, and vice-versa, goes through the brainstem. In the first place, it contains sensory pathways through which information pertaining to pain, temperature, touch and proprioception (perception of one's own body in space) pass [p96]. Secondly, it transmits the axon sheaths [p17] that originate from the brainstem motor neurons and end up in a spinal cord synapse, where the information that regulates movement is transmitted.

But that's not the end of it. From the three brainstem components (**medulla oblongata**, **pons** and **midbrain**) ten of the twelve pairs of **cranial nerves** emerge – these nerves originate directly in the brain, in both hemispheres, and are in charge of the motor and sensory control of the face and eyes, and also regulate other functions, including digestion.

The entire brainstem is criss-crossed by a **reticular formation**, a collection of a hundred or so neural networks that maintain a waking state, sending the cortex a sequence of signals at regular intervals. Whenever the sequence slows down, sleepiness is triggered. The reticular formation also plays a key role in attention mechanisms [p157]. If we add that, besides being responsible for breathing and the beating of the heart, the brainstem contributes to the development of knowledge and awareness – and, therefore, consciousness – it becomes obvious that it is at the root of everything. In every sense.

The medulla oblongata

You can't live without this component. Involuntarily – in other words without the slightest act of will on the part of the legitimate owner – the medulla oblongata rules life's basics, such as breathing, heartbeat and blood pressure. It's the liaison officer between the spine and the brain.

Since it's part of the brainstem, it conveys neural messages between the central and peripheral nervous systems. In the case of the latter, it plays a crucial role in regulating basic, unconscious respiratory and cardiovascular functions, as well as many automatic reflexes, involuntary responses to a number of stimuli (estimated at 45). Among these, we have coughing, sneezing, vomiting and yawning.

The pons

This is located along the brainstem, between the medulla and the midbrain. Also known as pons Varolii ('bridge of Varolius') – in memory of Costanzo Varolio, the sixteenth-century Italian anatomist who first described it – it has a slightly bulging shape.

The pons connects various areas of the brain and manages the vital flow of information between the cerebral cortex and the cerebellum. Crossed by four pairs of cranial nerves, it supports sensory functions such as hearing, taste, touch and balance, and also motor functions such as chewing or moving the eyes. Since REM sleep originates here, the pons even plays a key role in dreams.

The midbrain

Barely two centimetres long, the midbrain is the final section of the brainstem and also the smallest.

By the 28th day of embryonic development, give or take, the neural tube splits into three vesicles that are getting ready to become a brain. They are called the **hindbrain**, the **midbrain** and the **forebrain**. Within a few weeks, the hindbrain divides and will eventually form the pons, cerebellum and medulla oblongata. The forebrain will divide into the **telencephalon** (cerebral cortex) and the **diencephalon** (hypothalamus, thalamus, etc.)

Thanks to its strategic location, the midbrain is crossed by portions of white matter that connect the pons and the thalamus – in other words the 'reptilian' brain and the limbic system – in both directions. In addition to this, it stores a variety of grey matter nuclei, sensory as well as motor, which regulate waking, pain and hearing, as well as head and eye movement. It can be divided into three parts: the **tectum**, the **tegmentum** and the **peduncles**. The first two are separated by a central aqueduct (where the cerebrospinal fluid flows [p33]), while the peduncles are divided by the paired *substantia nigra*, one of the main sources of dopamine.

Substantia nigra

It's called 'black substance' because it is densely populated by neurons darkened by melanin, the same pigment that produces a sun tan. In this case, there are two black substances: the *pars compacta* is populated by dopaminergic neurons that project themselves into the nearby striatum (critical in both reward and motor systems), while the *pars reticulata* is mainly composed of GABAergic neurons [p23] that are connected to many different structures.

The ventral tegmental area

Still in the area where the midbrain ends, there is a tiny neuronal structure that is strategically connected to many different corners of the brain, from the brainstem to the prefrontal cortex. An essential part of the dopaminergic system and therefore of the reward system [p141], the ventral tegmental area (VTA) fuels motivation [p154], learning [p162] and, in a totally different category, orgasm [p119]. It is also connected to drug addiction and some serious mental illnesses.

3.1.2 The Cerebellum

In comparison with the cerebral cortex – the external, dominating part of the human brain – the cerebellum ('little brain') can seem like an accessory or the vestige of a distant evolutionary past. After all, it represents just 10% of the entire brain volume. Yet the role of the cerebellum should not be downplayed.

Only slightly bigger than a golf ball, it looks a bit like a small-scale encephalon – it too has two hemispheres. It lives in the rear cellar of your brain, beneath the **temporal lobes**, and is directly connected to the spinal cord, which transmits the brain's messages to the rest of the body.

The cerebellum is a part of the brain equipment of all vertebrates – reptiles, fish, birds and mammals. We've known for a few centuries that it deals with motor control, balance and movement coordination, since we've seen what happens when it is physically damaged. In the *Homo sapiens* model, however, evolution seems to have increased its features and importance. Recent studies suggest that this ancient part of the brain could have a far from minor role.

It is populated by about 69 billion neurons, in comparison with about 20 in the cortex. The secret to its diminutive size lies in the fact that about 46 billion of these are granule cells, some of the smallest neurons in existence. This tiny power-house deals with motor learning, especially when it comes to learning and performing sophisticated movements, such as a backhand half-volley on the tennis court or playing Bach's fugues on a piano. Its constant exchange of information with the cerebral cortex, almost in tandem, also suggests a role in cognitive functions, as well as the motor function for which it originally evolved.

3.2 THE 'MAMMAL' BRAIN

The limbic system is situated above the brainstem and beneath the cortex, and is made up of a vast number of small, intercon-nected structures. Although it is also present in invertebrates, it has grown larger and more prominent with evolution in the mammal brain. All its structures are replicated in the left and right hemispheres, sometimes with partially different func-tions. Every thalamus, amygdala and hippocampus has its mirror image. The hypothalamus is the exception.

The limbic system was long considered the 'emotional brain', but we now know that it is much more complex than that. Emotional experiences such as fear and love do largely depend on it, but so, too, do functions such as learning, moti-vation and memory.

It's thanks to the limbic system that you enjoy food and sex. And it's the limbic system's fault if you experience depression or lack of motivation. It's because of the limbic system that chronic stress could cause you to have high blood pressure. It's the limbic system that allows you – for better or worse – to tell someone to get lost.

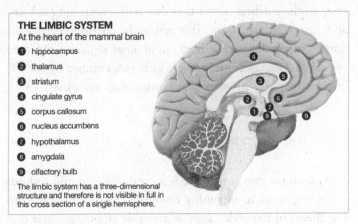

THE LIMBIC SYSTEM
At the heart of the mammal brain

1 hippocampus
2 thalamus
3 striatum
4 cingulate gyrus
5 corpus callosum
6 nucleus accumbens
7 hypothalamus
8 amygdala
9 olfactory bulb

The limbic system has a three-dimensional structure and therefore is not visible in full in this cross section of a single hemisphere.

3.2.1 The Thalami

Your brain needs an efficient system to sort through the flow of information constantly flooding into it, and to send this data to be computed in the appropriate areas of the cortex. This sorting office is made up of the two thalami. These are a pair of symmetrical structures, each the size of a walnut but more elongated, positioned almost in the centre of the brain and connected by a tiny strip of grey matter. Their job is of such vital importance that serious damage to the thalami causes irreversible coma. With the exception of the sense of smell [p97], all sensory modalities [p96] go through here, including proprioception, as does information coming from the largest sensory organ in the human body: the skin.

To give an example, the image received by the retina of the right eye is transferred to the left thalamus which, in turn, sends it to the **left occipital lobe** [p58], in other words the slice of cortex dedicated to vision. But the thalami aren't merely postmen, since they in return receive information from the occipital lobe. This mechanism is replicated in all the other areas of the cerebral cortex that deal with computing sensory and motor information. This results in an impressive thalamus–cortex–thalamus closed circuit that regulates the state of wakefulness and attention, which is a component of the 'magical' cerebral network that produces consciousness [p126].

3.2.2 The Amygdalae

The job of the two amygdalae is to find – within milliseconds – the responses to incoming emotions, and then to memorise them. In the wide range of available emotions, their true specialism is fear, an experience so important to survival that a special circuit has evolved just to manage it [p116].

The two amygdalae – enclosed in the right and left temporal lobes – work and memorise in tandem, but also appear to have a few personal preferences. While the right one is dedicated to fear and negative feelings (as we can confirm by electrically stimulating it), the left one is more open to positive feelings and is probably involved in the reward system. Each amygdala receives information from neurons assigned to vision, smell, hearing or pain, and resends executive orders to the motor apparatus or the circulatory system. For example, in case of danger, it simultaneously orders the body to keep still, the heart to beat faster, and the stress hormones to do their job [p196].

The amygdalae – which owe their Greek name to their vague resemblance to almonds – also manage memories of fear, including conditioned reflexes associated with it [p60]. A mouse whose amygdalae have been removed has no wish to run away from a cat. In addition, the amygdalae take part in the process of consolidating all long-term memories [p64].

Thanks to new brain imaging technologies [p234], it's become increasingly clear that a faulty performance on the part of the amygdalae – whether for genetic reasons or a neurotransmission error – can be connected with anxiety, autism, depression, phobias and post-traumatic stress [p184]. A trauma caused by war or sexual violence *physically* changes the amygdalae within a brief space of time. It is probably the most sexually dimorphic brain structure there is, in other words the one that distinguishes most the male from the female brain types [p178].

3.2.3 The Hippocampi

After the frontal cortex, the two hippocampi are probably the most scientifically controversial parts of the brain. Just to give you an idea, *The Hippocampus Book*, a work of neuroscience published by Oxford University Press, is 840 pages long. The hippocampi are shaped vaguely like seahorses – hence the name given them by the sixteenth-century anatomist Giulio Cesare Aranzi – and are located one in each hemisphere, between the thalami and the temporal lobes of the cortex. They deal with memory and space.

They are responsible for the installation of episodic memories, i.e. memories of personal experiences [p64]. They also play a part in semantic memory (relating to meaning in language), which can include ordinary notions as well as complex social rules. Finally, they are instrumental in consolidating memory, both short and long term. Any damage to the hippocampi makes it impossible to form new memories, while leaving old ones intact (since these are stored elsewhere in the brain) and preserving explicit memory, i.e. the ability to learn new manual skills (also located in another part of the brain).

The hippocampi are modulated by serotonin, dopamine and noradrenaline neurotransmitter systems [p20]. Yet scientists have noticed that they are also crossed by mysterious electrical impulses every 6-10 seconds, called theta waves (at a frequency between 6 and 10 hertz) [p13]. According to recent studies at Berkeley, it seems that these theta waves in turn transmit information – as proved by means of electrodes implanted in the brain of a mouse trying to find its way through a maze.

As a matter of fact, navigation is another core business of the hippocampi. This is shown by a famous study of the brains of London cab drivers, who have to memorise the huge capital city's map, and thus, as if by magic, have an enlarged hippocampus (the rear part).

These two cornerstones of the limbic system also contain a high quantity of cortisol receptors [p27], making them especially vulnerable to long-term stress. It has been proved that the hippocampi of people affected by post-traumatic stress are partially atrophied [p197]. There also appears to be a connection with severe depression and schizophrenia [p201].

3.2.4 The Hypothalamus

For something so small, this has a colossal task: to ensure survival. The hypothalamus, a 4-gram, 4-millimetre-thick structure buried in the middle of the brain, collects the most disparate pieces of information that arrive from the body. And, when the need arises, it turns on the chemical and neural switches to ensure the maintenance of homeostasis, i.e. the appropriate balance of essential resources.

The hypothalamus is located where the two hemispheres meet, low down in front of the two thalami. It has a right- and left-hand structure but, unlike the thalami, looks like a single object and is therefore considered as such.

By means of its numerous nuclei – the operating units of which it is made – it controls body temperature, measures the absorption of water and food through thirst and hunger, manages the constant physiological flow called the **circadian rhythm** [p111] and regulates sexual behaviour.

The tiny hypothalamus is so powerful because, besides its neuronal arsenal, it controls the **hypophysis** (or pituitary gland), the queen of the endocrine system, which lies nearby. The hypophysis produces eight hormones essential to homeostasis, two of which are synthesised by the hypothalamus itself. From the strategic growth hormone (which stimulates cellular reproduction and regeneration) to the corticotropin-releasing hormone (needed to fight stress). From oxytocin to vasopressin (two neurotransmitters essential for falling in love) to prolactin (which regulates lactation) and gonadotropin (which manages sexual development).

In other words, to the hypothalamus is assigned not just survival but the entire continuation of the species.

3.2.5 The Basal Ganglia

Deep in each of the two brain hemispheres, there's a collection of grey matter nuclei that are as well connected with the upper floors of the cortex as with the ground floor of the brainstem. They are associated with voluntary and automatic movement, eye movement, and also emotions and cognition.*

---

* The basal ganglia actually include brain components that derive from the telencephalon, diencephalon and midbrain, in other words three of the five subdivisions of the brain in embryonic form [p40].

The putamen

These are large, rounded structures that reside on top of the thalami and are involved in the complex motor mechanism. The putamen is linked to degenerative illnesses such as Parkinson's, which affects the motor system.

The caudate

This originates in the putamen of both hemispheres, and wraps them in a kind of tapering spiral. It, too, is involved in the motor system and is also implicated in Parkinson's. It performs cognitive (learning, memory, language) and psychological tasks: fMRI scans show that the caudate nucleus 'lights up' when you see your loved one, and also when you witness beauty in general. Together with the putamen, it constitutes the **dorsal striatum.**

The nucleus accumbens

Here we should really say *nuclei accumbentes*, since there is one of these round structures in each hemisphere. They are involved in the reward system [p141], being an essential component of the so-called mesolimbic pathway, which carries dopamine [p25] from the tegmental area [p41]. As such, they are part of the process of drug addiction. It has also recently been discovered that the *nucleus accumbens* is active in repulsion, the opposite of reward. It plays a part in impulsiveness and in the placebo effect [p188]. Together with the olfactory tubercle, it constitutes the so-called **ventral striatum**.

The striatum

The ventral and dorsal striata (the sum of the components described above) form the striatum, the entirety of which is associated with supporting learning and other cognitive functions, as well as the reward system. In general, the striatum is activated whenever you have a pleasant experience or even if you simply expect to have one [p60].

The globus pallidus
This receives information from the striatum and dispatches it to the *substantia nigra*. It plays a key role in voluntary movements.

The subthalamus
This receives input from the striatum and helps modulate movement.

3.2.6 The Cingulate Cortex

In our survey, we are midway between the mammal and primate brains. The cingulate cortex – enveloping the corpus callosum [p35] in both hemispheres – is part of the cerebral cortex, but, since it is considered an integral part of the limbic system, we include it here.

From the point of view of brain architecture, we could say that the cingulate cortex is the equivalent of the attic in the limbic apparatus. Since it receives information from above (the cortex) and below (the thalamus) [p43], the cingulate cortex is involved in emotions [p114], learning [p162] and memory. It performs a long list of vital roles.

The front part contributes to basic functions (such as blood pressure and heartbeat) as well as complex ones, including emotional control [p174], prediction [p60], and decision-making [p171]). The rear part, on the other hand, is a key piece in the **default mode network** [p167], but is also involved in retrieving memories and in consciousness [p126].

3.3 THE 'PRIMATE' BRAIN

The construction of the brain's upper floors began in mammals hundreds of millions of years ago. The telencephalon (what is considered the final section of the brain in its embryonic form [p40]) has progressively evolved in the cerebral cortex, the most powerful and sophisticated component in the brains of mice, cats, monkeys and the humans currently living. Reptiles and birds have a pallium, which some call a cortex, although it actually isn't one.

Over the course of entire geological eras, it was the development of the cortex that singled out hominids from other primates. The genus *Homo* appeared roughly two million years ago, and the now 200,000-year-old cortex of *Homo sapiens* – the only surviving *Homo* species after the extinction of the Neanderthals – became huge in comparison to the cortex of the rest. It's thought that contributing factors to this included the manual dexterity of their opposable thumb, the predatory abilities of their frontal and stereoscopic vision, as well as the social opportunities provided by a primitive language. The cortex continued to develop until the advent, about 50,000 years ago, of the culture and behaviour of the 'modern' human being, *Homo sapiens sapiens*. Just to give you an idea of its size, your cortex represents almost 90% of your entire brain weight.

This is where the brain's fireworks are lit. This is where the information chaos that floods in from connected peripherals, such as the skin and the eyes, is processed and catalogued. This is where new memories sprout and take root, before being classified and associated with already accumulated knowledge. It's thanks to the enormous computational power of the cortex that you are able to think, imagine, compare, decide and change your mind.

The cerebral cortex is the triumph of grey matter, i.e. the typical brain material packed with neurons, glial cells and capillaries, with that pinkish-grey colour observed by history's early anatomists. Grey matter is well distinguished from white matter, which is found here and there in the brain, but particularly below the cortex, where we meet the corpus callosum, hundreds of millions of axons linking the two cortex hemispheres together with white myelin [pp17, 32].

Present only in the brains of placental mammals, the corpus callosum is an integral, indispensable part of the most complex brain mechanisms, such as intelligence and consciousness, because it turbo-charges the two hemispheres of the cortex.

3.3.1 The Cerebral Cortex

If you took a tablecloth measuring $2m^2$, spread it on the table, then squeezed towards the middle, as though trying to fit it into a vase or a box, you would obtain something convoluted and wrinkled that looks very similar to a cerebral cortex. And that's precisely what evolution has done. It has organised the grey matter of the neocortex – the most extensive and evolutionarily youngest part of your brain – in a way that maximises the available space inside an increasingly large cranial box.

The valleys, the concave parts of the tablecloth, are called **sulci**. The mountains, or convex parts, are called **gyri**. The **fissures** are the deeper furrows that distinguish the four lobes of the cortex, each in turn named after the corresponding cranial bone: the **frontal lobe** in the front (location of abstract thinking, reasoning and also social skills and personality), the **temporal lobe** on the side (hearing, comprehension, language and learning), the **parietal lobe** at the top of the back of the cranium (touch, taste and temperature) and the **occipital**

lobe above the nape of the neck (vision). Finally, there is a deep division called the **longitudinal fissure**, which separates the cortex into two hemispheres.

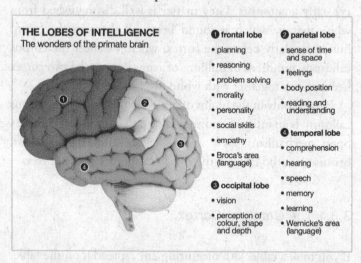

THE LOBES OF INTELLIGENCE
The wonders of the primate brain

❶ frontal lobe
- planning
- reasoning
- problem solving
- morality
- personality
- social skills
- empathy
- Broca's area (language)

❸ occipital lobe
- vision
- perception of colour, shape and depth

❷ parietal lobe
- sense of time and space
- feelings
- body position
- reading and understanding

❹ temporal lobe
- comprehension
- hearing
- speech
- memory
- learning
- Wernicke's area (language)

It was believed for a time that the two hemispheres performed different tasks and that every user had a dominant – let's call it favourite – hemisphere. That's how the legend was born about people with a 'left brain' being good at maths and logic, and those with a 'right brain' being more creative and artistic. This is not true [p206].

As a matter of fact, following a hemispherectomy (the surgical removal or deactivation of a hemisphere, carried out in cases of rare epilepsy syndromes), the brain and cognitive processes often resume working in time, especially if the patient is a child and therefore capable of manifesting full neuroplasticity [p69]. There are a very small number of users in the world who lead a more or less normal life with a single hemisphere.

This doesn't mean that the two hemispheres do not present particularities – this is what neuroscientists call laterality. The most famous examples are **Broca's area** (language production) and **Wernicke's area** (language understanding), usually located in the left hemisphere, but which are in the right hemisphere in the case of some left-handed individuals. Let's not forget that the data arrives from sensory peripherals, which flood into both hemispheres but with a reverse modality: information from the right eye is processed by the left occipital lobe, and the sense of touch in the left hand is computed by the right parietal lobe.

The 'tablecloth' of the cerebral cortex is between 2 and 4.5 millimetres thick. And yet this thin, squishy grey matter is made up of no less than **six neural strata**, each with its own structural characteristics, both because of the various types of neurons it contains and because of its connections with other cortical and subcortical (i.e. under the cortex) zones.

The cerebral cortex is the most complex structure in the brain, which, in turn, is the most complex structure we know of. And, as technological progress allows science to scrutinise the brain mechanism increasingly closely, there is evidence that no two cerebral cortexes in the world are the same, or work in exactly the same way.

This is extraordinary, since only one tiny variation of 0.1% in the genome determines the difference between an Australian Aborigine and a Greenlandic Inuit, so they are practically identical genetically. And yet, 100% of *sapiens* brains are unique.

The frontal lobes
Welcome to the central control station. Here, in the frontal cortex – the prefrontal cortex in particular, which occupies the forehead – is a concentration of your most sophisticated cognitive functions, such as thought and reasoning, beliefs and behaviour. This is what truly distinguishes your *sapiens*

brain from previous versions: the prefrontal cortex is much more developed than in most other primates, and doesn't seem to exist at all in many other mammals.

In human beings, the frontal lobes reach full operational mode 25–30 years following the start-up of the brain [p79]. This largely explains the marked differences between childhood, puberty, adolescence and adulthood.

It is impossible to overestimate their role. Just to give you an idea, we invite you to do four successive exercises:

- Choose a holiday location from your childhood and picture it as it used to be, in as much detail as possible. Then imagine what it would be like today.
- Try to imagine, in an armed conflict situation of your choosing, the hard life of a woman left alone with two young children.
- Start from 101 and count backwards by subtracting 8 at a time.
- Put your hand on the table and start drumming with your fingers in both directions.

Good. Once you've finished, you will have activated the frontal lobes of your brain entirely. You would never have been able to perform these tasks without a *sapiens* cerebral cortex. It's in the frontal cortex that we can recall an emotional memory mediated by the limbic system, then shape it by means of imagination [p167]. This is where empathy [p132] towards other living beings comes from. This is where calculations, logical reasoning and language are managed. But it is also here that voluntary movements, such as finger articulation on a computer keyboard, are regulated, and here that the primary motor cortex resides, allowing you – among other things – to walk.

The frontal lobes are the control panel because they manage the so-called executive functions of the brain, such as **operative memory, inhibitory control** (the ability to react differently from normal in order to reach a goal), **delayed gratification** (the ability to resist a desire in exchange for a future reward), **cognitive flexibility** (the ability to manage several concepts at once), **reasoning**, **planning** and much more [p153]. Our personality is mainly here, in the prefrontal cortex [p149].

We've known this for a long time thanks to a dramatic incident. Before the advent of non-invasive imaging technology, such as PET, MEG and fMRI [p234], scientific conclusions often used to be reached by physically comparing the brain before and after surgery, strokes or violent trauma. This manual deliberately avoids detailing such gruesome research methods, but cannot resist the most famous case study.

In 1848, in Vermont, Mr Phineas Gage was working on the building site of a new railway. There was a sudden explosion and an iron rod, a metre long and 3 centimetres in diameter, pierced Mr Gage's brain from bottom to top, perforating his left frontal lobe. Incredibly, the poor man survived. However, he was no longer the same Phineas Gage he once had been. Whereas before he had been kind and beyond reproach, he had now become dissolute, irreverent and a womaniser. 'The balance between his intellectual faculties and his animal tendencies,' John Harlow, the doctor who described this clinical case summarises, 'seems to have gone haywire.' The Gage case has a place in medical history because it proved beyond the shadow of a doubt that biology and psychology are closely connected.

We strongly recommend that all users, especially those who carry their brain on a motorbike or skis, take precautions so as not to traumatise the central control panel unit.

The temporal lobes

On either side of the primate brain, located at ear level, are the temporal lobes. Assigned primarily to language and sensory perception, they process sound signals from the auditory peripherals through the so-called primary auditory cortex [p106], which is, in turn, connected to secondary processing areas where sounds and words are interpreted. Right next door, but only in the left temporal lobe, is Wernicke's area, which specialises in the comprehension of language, both written and oral. Research has shown that, while lesions in this area of the cortex don't prevent a patient from being perfectly capable of talking (because Broca's, in the frontal lobe, takes care of that), they do impair the ability to put words in order.

As is the case in many other areas of the cortex, the capacity of the primary auditory cortex is closely linked to experiences – in this case auditory – in youth and childhood. This is why a small child who is exposed daily to the sounds of two or three languages will grow up to be a polyglot. If he hears these languages only during his years at kindergarten, he will continue to make out the sounds of this language, even if it doesn't become his mother tongue. For the same reason, a little girl who has never been exposed to music during the first ten years of her life is very unlikely to become a professional musician. There are even those who theorise that listening experiences as early as the foetal stage (with headphones put against the mother's belly) predispose a future baby to being musical. A video was posted on YouTube in 2015 of Dylan, a little boy who was given music to listen to starting from five months before his birth, and who is able to hear a chord – even a complex one – and immediately identify the notes that form it.

The temporal lobes are also essential to two other

fundamental functions: vision and memory. In the first case, they receive visual information from the occipital lobes and decode it, associating every detail, like faces and objects, with their names [p102]. In the second, the temporal lobes communicate with the hippocampi and amygdalae in order to form long-term memories [p64].

The parietal lobes

It's eight o'clock, time for breakfast. Picture the following, perfectly normal scene, in slow motion: your hand reaches out for a cup of tea but pulls away because it's boiling hot. So it immediately alters its plan and carefully lifts the cup by the handle instead.

What's so special about this? Quite a lot. To start with, this ordinary, carefree action requires a visual system capable of processing information about the distance from the cup to the hand, as well as its shape and characteristics. Next, the movement is accomplished with the caution necessary for checking the external temperature: a system is needed which is capable of computing data sent by the hand's receptors, and also then converting the strategy into another spatial-sensory calculation focused on the handle of the cup. Put simply, we need two parietal lobes.

Located just above the back of the cranium, right behind the prominent frontal lobes, the parietal lobes deal with multi-sensorial perception and its integration into the motor system. In this sense, they could be considered an 'associate cortex' that processes signals from all available senses – vision, hearing, thermoreception, nociception and so forth [p96] – and here serve the specific purpose of saving you from scalding yourself each time you drink a cup of tea.

Along the entire border that separates the parietal lobes from the frontal lobes, there's an area called the **somatosensory**

cortex, which controls touch. This also contains a spatial-sensory map that allows it to discern that the pain of the hot cup came specifically from the hand. This neural representation of all the tactile regions is called the **homunculus** (little man) and it is proportionate to the number of receptors in the more sensitive parts of the body: imagine a humanoid figure with enormous hands and feet, as well as a huge tongue [p109]. The parietal lobes are also largely involved in language and its decoding.

The occipital lobes

Just as the left ear sends the right temporal lobe (on the opposite side of the brain) acoustic signals converted into electrical signals, the left eye sends the right occipital lobe (on the opposite side of the head) light signals converted into electrical signals.

Yet it would be an understatement to say that the occipital lobes simply deal with vision. There's an area in the occipital lobe that receives visual information and another that interprets it. It's their work in tandem that allows you to read these words and instantly understand them.

First and foremost, each lobe must receive the enormous amount of data (and this is upside-down data, on top of everything else) that comes from the retina of the opposite eye through the thalamus. Then it must simultaneously calculate the colours of all the objects in the field of vision, estimate their size, distance and depth and identify moving objects or familiar faces. This complex task is performed by various areas of the cortex, which work side by side.

After the **primary visual cortex** (known as V1) receives the crude information and detects movement, other areas deal, for example, with making associations (V2), detecting colours (V4) and computing the shape, size and rotation of the objects.

The sum of these operations produces the 120-degree, colour, three-dimensional image – largely high resolution and in real time – that your brain represents to you at that moment [p102].

But things don't end there, because the occipital lobe in turn forwards this information from the field of vision to the parietal lobe – so you can to perform tasks such as picking up a cup – and to the temporal lobe – so as to connect current visual information to memories of past data (including remembering that the cup could be scalding hot).

4.0 MAIN FEATURES

YOUR BRAIN IS A MULTITASKING DEVICE. Three in one? Five in one? No, it performs many more tasks than that – so numerous and so interwoven that it would be difficult, if not impossible, to list them all. Your brain can think and react, remember and forget, fall in love and hate, sleep and wake up, understand and learn, build and destroy. And what else? After all, its power is the power of humanity.

If, however, we were to pick three or four of its main characteristics, they might be these: that the brain is constantly reorganising its connections, forever trying to foresee the future, can think about itself, remember, and feel the weight of its own existence.

The sum of all this is called intelligence.

4.1 PREDICTION

While you're living more or less serenely in the present, your brain is constantly busy imagining the future. Numerous neurological and psychological experiments have proved the existence of a basic brain function which our ancestors – who lacked technology such as magnetic resonance – could never have imagined: prediction. In a nutshell, the brain is constantly predicting its own perceptions, always looking into the more or less distant future.

You're not aware of this, but while you walk, your brain is predicting at every step when your foot will reach the ground. If the prediction fails – because there are stairs or a small hole – you know exactly what's happening and an *instantaneous* state of alarm prompts you to retrieve your balance.

If we could not anticipate the trajectory of cars, we could not sit behind the wheel – even crossing the road would be dangerous. Many sports would be impossible, since we would be unable to calculate the direction of a ball or foresee where gravity would make it land. Once upon a time, before playlists and streaming broke up the traditional listening order, every Beatles fan would experience a peculiar effect: after listening to 'With a Little Help from My Friends', they would, during that brief pause, already hear in their heads the first notes of 'Lucy in the Sky with Diamonds', even before the record began to play the song.

Your brain is constantly busy imagining the future in order to compare it to the past. In the chaos of signals it receives, it uses old experiences to anticipate the perceptions which will, presumably, arrive straight afterwards. When the forecast turns out to be wrong, as in the case of a wrong step, it draws on past experiences to correct the error.

Not so long ago, science corrected one of its own errors. For centuries its was believed that the brain reacted to information that reached it from the sensory organs. However, we now know that the brain does not react, but predicts. That's why it's constantly active, and crossed by millions of chemical reactions each second. The brain predicts senses and sensations – for example visual, auditory, olfactory – and compares them to past experience without your noticing – at least, until an unexpected signal calls for your attention. Anyone used to driving their car down a familiar stretch knows what it's like for their mind to wander until the next red light.

Even when you're paying attention, when listening to someone speaking, for instance, your brain automatically selects sounds, syllables and words, trying to anticipate the sounds, syllables and words – and consequently the concepts – that will follow. The same thing occurs if you see an old film again (foreseeing which scene comes next) or when you watch it for the first time (imagining how it will turn out).

Prediction is closely linked to the reward system [p141], one of the most important neuronal mechanisms that influence behaviour, which it achieves by activating the dopamine circuit [p25]. At the beginning of the 20th century, Ivan Pavlov discovered so-called **classical conditioning**: by making the same sound every time he fed his dogs, the Russian psychologist noticed that soon all it would take was to play the sound and they would automatically start to salivate. Pavlov had no idea that dopamine circuits were involved (these would be discovered later, in 1958), but his experiments laid the foundations for studying cognitive processes.

Studies on monkeys have proved that, once the mechanism for obtaining food has been learned (for example pushing a button five times), the primates receive a discharge of dopamine which gives the brain a feeling of pleasure. After some time, the flow of dopamine no longer arrives together with the food, or when the monkey presses the button. It comes much earlier, when the monkey – delighted to have discovered the trick that will fill its belly – is *about* to push the button. It's an anticipating mechanism. Contrary to what was once believed, the reward doesn't come at the end, but *before* the action is performed. And so we discover that neural prediction, this projection of the brain into the future, fuels motivation. Dopamine floods the brain in order to trigger the action *beforehand*, and not to reward it after the event.

It's interesting to observe that when researchers started

to reward the monkeys only 50% of the time, discharge of dopamine was not halved, but was more than doubled. In other words, when faced with a certain degree of uncertainty, the reward system actually multiplies the feeling of wellbeing modulated by the dopamine. Perhaps it's this basic inconsistency that encourages many owners of operating brains to play at roulette (one chance of winning in 38) or play Lotto (one chance in 45,057,474), sometimes to the point of being unable to give up [p193].

Looking at it another way, one could say that the brain has this obsession with the future because it's the only method it possesses to manage life's unforeseen events. This is an evolutionary motivation. The overwhelming chaos of internal and external information [p96] that your brain processes every second is often somewhat unclear and ambiguous, so the brain makes up for it by trying to imagine what is going to happen. It must make a huge sequence of inferences (from the Latin *inferre*, 'to bring inside') to predict the immediate future.

Bayesian inference, named after the eighteenth-century priest and mathematician Thomas Bayes, is based on a statistical theorem that estimates the variation of probabilities of an event, depending on the variation of available information ('the probability of A given B equals the probability of B given A, multiplied by the probability of A and divided by the probability of B'). The complex mathematical statistics that originated with Bayesian inference has been applied to engineering, medicine and philosophy. **Computational neuroscience**, which studies brain function in terms of data elaboration, considers the brain a Bayesian machine that produces constant inferences about the world and adjusts them on the basis of actual sensory perceptions. This approach is now having important consequences in the development of artificial intelligence [p242].

The predictive properties of the brain could provide the key to understanding human intelligence. Jeff Hawkins, who invented the Palm Pilot, a pocket-sized computer in the '90s, established an artificial intelligence start-up based precisely on these neurological mechanisms. In his book *On Intelligence*, Hawkins defines intelligence as 'the capacity to remember and predict patterns in the world [...] by combining what has been seen before and what is happening now'. This actually underplays things. 'New scientific studies,' writes Lisa Feldman Barrett, psychologist at Northeastern University, 'suggest that thoughts, emotions, perceptions, memories, decisions, classifications, imagination and many other mental phenomena considered historically as distinct mental processes can be brought together under a single mechanism: prediction.'

Prediction, a brain function that's totally imperceptible to the user, requires another important mechanism in order to be activated correctly. So that it can contrast the uncertain future with a known past, the brain needs an incorporated memory.

4.2 MEMORY

You are what your brain remembers. Without memory, you wouldn't be able to speak, move in space, have social interactions and consequently be what you are. You would be deprived of your personality.

We are all made up of what our ancestors have bequeathed throughout history. Without memory, human civilisation and the social groups we know would not exist. The memorisation of language has allowed the creation and flow of cultures, with streams of oral traditions, rivers of books and, nowadays, oceans of multimedia information.

Your installed memory is 100% compatible with the current

system version of the brain [p10]. This is a definitively human piece of equipment. Memory first developed as a fear mechanism, a reminder to keep away from danger. Vertebrates also acquired spatial memory to improve their navigation through the world – useful both for prey and predators. In mammals, a kind of social memory evolved, complete with hierarchy and family relationships. In primates, a motor memory has grown. And in humans, subjective memory was added, which marks out personality, in all its wide range of colours and shades, and allows it to be projected into society [p149].

There is no central storeroom for information. Instead, it is distributed in such a complex, intricate synaptic network that we have yet to comprehend most of it. Every fragment of memory (words, panoramas, emotions) is encoded in the area that created it (temporal lobe, occipital lobe, limbic system) and reactivated every time it's brought back to mind.

Not only is memory not a uniform process, but there are multiple types of memory, each encoded in different areas.

Short-term memory is truly short: just a couple of dozen seconds. It's as though it's constantly recording our life's events, such as the things, people and shop windows we come across while taking a walk through the historical centre of a city. If all this information isn't recalled by means of associations or a concerted effort to memorise it, it vanishes for ever shortly afterwards. So long, that is, as there is no hyperthymesia, a very rare disease that forces a sufferer to remember in every detail how they were dressed on 13 April 2005 and the words they uttered before breakfast. Usually, we humans can barely remember a phone number we heard a few seconds earlier. **Operative memory** is part of short-term memory. It's what we use, for example, when we mentally repeat the phone number in order to store it for a few more seconds.

Short-term memory is essential for anchoring **long-term**

memory, which is, in essence, everything we know. It encompasses reconstructions of the most important events of our lives, a vocabulary of meanings (sometimes in several languages) and a catalogue of the most disparate manual and motor skills, in addition to names, numbers, faces, locations, events, concepts, emotions, sensations, qualities, opinions and beliefs.

For those who love classifications, it is usually divided into two:

- **explicit memory**, which can be **episodic** (the menu at the last Christmas dinner, Mum's date of birth) or **semantic** (Moscow is the capital of Russia, you need a ticket to go to the theatre).
- **implicit memory**, which concerns automatic motor memories (writing with a pen, riding a bicycle) and consequently also classical conditioning [p60]. We could also add **spatial memory**, which relates to spatial awareness, such as the ability to navigate an unfamiliar city.

Long-term memory includes recent events (the old friend you bumped into during this morning's walk), as well as very distant events (that summer holiday you once spent together). Joining together the friend and the holiday is an example of association, and human memory works mainly through **mechanisms of association**. It's very easy to remember an event when we connect it to something we already know. It's no wonder memory champions use associative strategies, connected, for example, to a familiar route through a known environment, to remember impossible sequences. In 2006, Akira Haraguchi, a 70-year-old Japanese engineer, recited by heart the first 100,000 decimals of the number Pi (3.1415926535 etc.), starting at 9 a.m. and finishing the following day at 1.28 a.m.

However, memory by association also produces something else: the multi-sensory reconstruction of events. Those two distant weeks at the beach seemed to have vanished from your memory until seeing your friend – the co-protagonist of that forgotten film – also brought back to your mind the heat of that summer, the scent of the pine grove, the fear of exams and the World Cup on TV. Still, the reconstruction of the past, conditioned by emotions, can also be altered and even totally wrong [p192].

Like any good archive, long-term memory needs to encode information, store it and know where to find it again. Although we don't know the precise biochemical translation mechanisms of information, this entire process is clearly connected with learning [p162] and, consequently, with consolidating synapses [pp18, 69].

Repetition is necessary for consolidating memory: your secondary school teacher used to say it and so does neuroscience [p162], but repetition is not enough. Without attention, i.e. if your brain isn't focused on what you're reading or listening to, repetition is of little use [p157]. Without motivation, the inner drive of curiosity or the emotional impact of a future reward (a university degree, for example), keeping your attention can be hard work [p154].

Memories tend to become well imprinted when they are associated with a strong emotional state, for example when they are linked to sad or happy episodes. Everybody remembers where they were and what they were doing when the news of the 11 September 2001 attacks broke. When the event is exceptionally traumatic, its memory can give rise to an unpleasant syndrome of post-traumatic stress [p196]. There are also cases where the memory of shocking events, especially if experienced in childhood, is involuntarily wiped out in its entirety.

Finally, as we mentioned previously, memory is **contextual**, in the sense that all the information is imprinted along with visual, auditory and sensory information experienced at the same time. In trying to remember an event or fact, it is therefore useful to remember its context: the mechanism of association often allows us to retrieve missing information.

How large is the memory installed in your brain? The answer to this significant question varies – there are those who declare the calculation 'impossible' and those who, like Professor Terry Sejnowski of the Salk Institute in California, compare the memory to the binary mathematics of computers and estimate it at one petabyte, or a noteworthy million gigabytes. Having said that, nobody's ever heard of anyone reaching 'the end' of their allotted memory storage.

Some say jokingly that the human brain is the only container in the world that can contain more liquid the more you pour into it. And yet that is exactly the case – for three reasons at least. Firstly, the associative mechanism of memory allows us to save space, avoiding many duplications. Secondly, speaking more than one language, learning to play a musical instrument and actively attempting to imprint new things on the memory not only slows down the ageing process of neurons [p224], but also gradually increases our learning ability: the more liquid you pour into the container, the easier it becomes to add more. Finally, learning, understanding, studying in depth, if not radically changing your opinion or position, truly adds something to the brain, since it physically changes it.

Memory depends on the brain's ability to remodel connections every second. It's an extraordinary mass-produced feature, pre-activated in your brain prior to purchase. It's called plasticity.

4.3 PLASTICITY

The Italian anatomist Michele Vincenzo Malacarne conducted a rather strange, macabre experiment. He raised two puppies from the same litter and a few pairs of birds from the same brood. Then, with a bit of patience, he spent two years training just one animal from each pair, leaving its sibling deprived of any kind of stimulus. Then he killed them all, and compared their respective brains to see if there were any differences between them.

The reason Malacarne was able to conduct this experiment – which would now be considered detestable, to say the least – was that he lived and worked in the eighteenth century. His experiment was destined to contribute a precious nugget of information to science: the animals that had received training and more stimuli had a visibly more developed cerebellum. Malacarne discovered that sensory experiences *physically* alter brain structure. It's a shame that, for almost two centuries, nobody took notice of his extraordinary finding, which contradicted the established concept that the physical brain was essentially unchanging.

The brain doesn't just change: it's constantly changing. All you have to do is watch a documentary, attend a conference or chat to your friends in a bar: every new piece of information, every new experience, every conclusion makes something move in the nanoscopic neuronal microcosm.

This extraordinary evolutionary trait is called plasticity. It's at the basis of the integrated memory and learning systems. Other animals have it, too, but among mammals, especially in *Homo sapiens*, it is amplified by the large cerebral cortex and the presence of language.

Plasticity adds new neuronal connections through axon

terminals on one side and dendrite branches and their spines on the other [p15]. Just to give you an idea, the number of spines, as well as their shape, can change within the space of minutes or even seconds. It can increase or decrease within a few hours in a frantic succession of old and new connections that modify, albeit slightly, the cabling of neural circuits.

But there's also synaptic plasticity, which strengthens or weakens connections between neurons. It was the Canadian scientist Donald Hebb who discovered that when two neurons are active at exactly the same time, the synapse that unites them grows stronger. As Hebb's rule states, neurons that fire together, wire together. It's thanks to Hebb's work that we've managed to discover synaptic strengthening mechanisms like long-term potential (LTP) [p12]. The opposite of the long-term strengthening of the memory is called **long-term depression**. This is a reduction in the efficiency of synapses, which is a precursor to forgetfulness.

Memory involves many areas of the brain: the basic structures for consolidating it are the temporal lobes of the cortex, the hippocampi and the connected structures of the limbic system. Consolidation comes through repetition. The hippocampus [p45], strengthened by its multiple synaptic ramifications, triages the pieces of information so that they can be associated accordingly.

American neuroanatomist James Papez considered the so-called Papez circuit, a 35-centimetre neural loop that leaves and returns to the hippocampus across the limbic system and the temporal lobe of the cortex [p51], essential to the mechanism of emotions. Nowadays we know that it is more essential, in fact, to the mechanism of memory: after the associations generated by an event or piece of infor-mation have taken a few high-speed trips along the Papez circuit, they become consolidated in the cortex. After that, they may not even require the hippocampus any longer. This explains why patients with a lesion in the right and left hippocampi are unable to form new memories but can perfectly well remember the distant past.

With plasticity, evolution has created a function indispensable to life as we know it. Neural circuits and synapses are relentlessly reorganising themselves to allow the brain to learn from everything surrounding it. This way, the brain is freed, at least in part, from the restrictions imposed by its own genome, which is carefully filed away in the nucleus of every one of its cells. We are at the exact crossroads of nature versus nurture. What is more important – the nature imposed by DNA or the culture absorbed by learning from the surrounding environment? It's a nagging question that carries philosophical and ethical implications. There are those who support the first and others the second. The good news is that we can simply reply – without fear of contradiction – that they are equally important – and it's lucky for us that we don't rely on just one.

In the not-too-distant past, people believed that development of the brain – after its tumultuous progress from the antenatal era to the age of three – gradually slowed down until it ground to a halt towards the end of adolescence. Nowadays, however, we know that in response to changes in behaviour, the environment, thoughts and emotions, the brain alters, imperceptibly but constantly. It has an innate ability to create new connections, reorganise neural pathways and, in extreme cases (following particular types of damage, for instance), even create new neurons.

The notion that personality, talent and abilities are static throughout a lifetime has proved entirely unfounded. Personality can be improved, talent developed and skills accrued. As we go on, unpleasant habits can be corrected and a new language can be learned. This capacity opens the doors to new horizons for the owners of an operating and intelligent brain such as yours.

- For voluntary functions, see the Control Panel section [p153].
- For correcting undesired plastic effects, see the Habits and Addictions section [p193].

4.4 INTELLIGENCE

Thanks to the combination of the memory of the past, the plasticity of the present and the anticipation of the future, intelligence emerges from the gelatinous *wetware* of the brain. This is the most grandiose and spectacular asset your *sapiens sapiens* central nervous system is equipped with.

To start with, it can be difficult to describe. If we summarise it by saying it's the ability to perceive our environment, process its information and store it, ready for any eventuality, then intelligence is not a human prerogative. Nor solely primate or exclusively mammal. The billions of years of natural selection has distributed various degrees of intelligence over the entire globe. It's just that the intelligence of primates and mammals has become superior.

Starting from genus *Homo* (two and a half million years ago), to the *sapiens* species (200,000 years ago), and on to the *sapiens sapiens* sub-species (50,000 years ago), social interaction, tools, language and, later on, writing, have raised the bar to produce da Vinci's imagination, Bach's inspiration and Hegel's rationality. It is the consecutive ability to communicate-understand-learn-invent that has marked an evolutionary difference, triggering a virtuous circle that gave birth to science, art, music and philosophy.

A definition of human intelligence could therefore include understanding, learning, self-awareness, creativity, logic and the ability to solve problems so as to adapt to increasingly complex circumstances. There are some who suggest that there is more than one type of intelligence – for example Daniel Goleman, with his **emotional intelligence** (the ability to read and interpret other people's emotions), and Howard Gardner, who claims there are as many as nine: **naturalist**, **musical**, **logical-mathematical**, **interpersonal** (corresponding to emotional intelligence), **intrapersonal** (the relationship with oneself), **linguistic**, **existential**, **bodily-kinaesthetic** and **spatial**. At this point we could discuss how consciousness [p126] is an integral part of the brain – only science and philosophy disagree so strongly on how to define both that we had better drop the subject.

Notwithstanding the impolite sense of supremacy towards the other animals on the planet that has distinguished *Homo sapiens sapiens*, the intelligence inherent to our genome has not in fact developed further over the past 50,000 years. Even so, from the silica of flint used to engrave tablets to the silicon of the processors that turn on smartphones, humankind has found ways of exponentially increasing available knowledge: when Gutenberg died, in 1468, there were between 160 and 180 Bibles around, but now about 10 million new web pages are published every day.

Intelligence has been the subject of fierce scientific debates, above all in relation to social mobility. Long associated exclusively with talent (a natural gift) and social class (heredity), the notion of static intelligence was consolidated further at the beginning of the 20th century, with the appearance of **intelligence quotient** (IQ)-measuring tests, often used for fuelling ethnic and racial prejudice. It's a shame, because when Alfred Binet, the psychologist who experimented with the first IQ test in French schools, began his work, his goal was to instruct teachers on how to help struggling young brains to learn more and better. He defined intelligence as common sense, judgement or the 'ability to adapt to circumstances' – something that could be used as a tool to improve people's position – rather than as a way to enforce social difference.

A century later, we have proof that Binet's attitude was correct: intelligence is not static, immovable and preordained. This is confirmed by the so-called Flynn effect, named after the scientist who discovered it (though he partially retracted his thesis later on): over the century, between the early tests and the present day, the population's average IQ has risen substantially. Are we more intelligent than our grandparents and great-grandparents? How can this be? Since our genetic heritage has not changed over such a short space of time, the answer to this riddle – besides justifying the odd doubt about the methods used for measuring intelligence – can only be in culture.

Long before the invention of agriculture, our hunter-gatherer ancestors were able to use a primitive language, learn from one another, and organise their tribal societies following a cooperative model, thanks to their prominent cerebral cortex [p51]. From their earliest years, the brains of a globalised modern society have at their disposal a wider variety of options for adding knowledge and creativity to their intellectual equipment. We are speaking here of software (your nanny's voice, your kindergarten experience, playing with friends) as well as hardware solutions (toys, books, computers, tablets, video games), which are capable of associating new modules with the installed version of your brain system. The brain produces culture, true. But it's also true that culture physically alters the brain.

It has been proved that when a brain user believes that

intelligence is a static, immovable cerebral property – a monolithic product of destiny – he or she can become a victim of the 'stereotype threat', i.e. involuntarily confirm myths about the intellectual inferiority of a particular race, social class or gender [p189]. On the other hand, countless psychological studies show that when the brain is convinced that it has no limits, then it can truly spread its wings.

Carol Dweck, psychologist and emeritus professor at Stanford University, has experimented in the field. Many children are convinced that intelligence and talent cannot be expanded – what she calls a '**fixed mindset**' – but when they are encouraged to move towards a '**growth mindset**', the educational results can be astonishing. According to Dweck, a fixed mindset thinks that intelligence, or talent, is 'set in stone' – it thinks 'you've either got it or you haven't' – and sees, more or less consciously, the effort of learning something new as pointless as well as tiring. By using more than one psychological lever, for instance swapping the pass/fail marking system for a 'you did make it'/'you didn't make it yet' system, you favour a more dynamic approach to learning. In other words, there is evidence that it is possible to change to a growth mindset [p162]. This way, the brain becomes more intelligent if it thinks it can become more intelligent. Obviously, this applies to all ages, not just childhood.

But what are the limits of intelligence? Will humankind be able to increase the intellectual power of the species without the lengthy process of natural evolution? Or has the evolution of intelligence observed on this planet – from the primordial nervous systems to abstract thought – reached the end of the line?

Evolution has produced numerous examples of intelligence, from dogs to mice, from dolphins to men, and it seems almost certain that humans will be able to reproduce this process.

Assuming that technological and scientific progress carries on at a good rate and for a long time, the construction, in the future, of a machine of 'human' intelligence level is practically inevitable. Some say it will happen by 2050, others before that. Even if it's ten years after that, the evolution of intelligence doesn't seem to be destined to stop with *Homo sapiens sapiens*, but to continue in the form of solid-state electronics. It's as if humanity is transferring the baton of intelligence to algorithms [p242].

Who knows? Given current scientific progress, it's not too absurd to imagine a future convergence between biological and digital intelligence. Something much more advanced than the present tendency towards augmented reality but going in the same direction, such as, for example, neural chips interfaced directly to the brain [p234]. In the meantime, given the giant steps made by technology in transcribing the genome, we will inevitably go searching for those few genes (1.2% of the total) that distinguish that of a human from that of a chimpanzee. Moreover, technological advances in genetic correction (such as the powerful CRISPR-cas9, DNA sequences which allow the editing of genetic information) herald possible future attempts to correct and improve the human genome [p238].

The evolution of intelligence definitely doesn't stop here.

5.0 INSTALLATION

Your brain is delivered to you pre-installed, so there is no need for complex connections or settings selections to make it work. It does, however, require some care during start-up, a process which takes the first few years of life, as well as maintenance in the longer term.

For it to work correctly, it's important from the outset to take good care of the quality of energy refuelling (commonly referred to as food), essential recovery and adjustment cycles (sleep), as well as the effectiveness of all the mechanical peripherals (physical exercise).

Let us remind you that your product is not covered by any guarantee [p248].

- For voluntary functions, see the Control Panel section [p153].
- For involuntary or semi-voluntary functions, see the section on Operation [p96].

5.1 PRIOR TO START-UP

It's both an extraordinary and ordinary event. Wonderful and mysterious. Practically perfect but imperfectible. It is the assembly of the brain, the nine-month process that precedes the start-up of the most beautiful and complicated machine in the world. Three weeks after work begins, before the maternal machine has even been informed of what is going on, a few stem cells already start to replicate and differentiate themselves. They are **ectoderm** cells, the most external of the layers, called germ layers, that constitute the tiny embryo. They are just about to specialise: they can become the progenitors of skin, tooth enamel or neurons. In the case of the last of these, the neuroectoderm gets reorganised in order to constitute the **neural tube**, which functions as the neuron factory.

The assembly line is already underway. The new neurons start to migrate towards their final destination, which, mysteriously, they know: even though they are inside an embryo of just a few centimetres, given their size (about 0.4 microns, or millionths of a metre), this is an extremely long journey. Once they have reached their destination, they start to assume the typical properties of that specific zone of the brain. They develop dendrites and axons, which will form the first synapses.

Two weeks later, neurons are born at the impressive rate of 250,000 new cells per minute. New connections are formed at the rate of millions per minute. And the migration of these new cells, which must cover increasing distances because the entire structure has grown, resembles a biblical diaspora. Despite this rush-hour traffic, every neuron knows exactly where it's going, what it's doing and what it's becoming; it's all written in the instructions for building the individual, instructions that are stored in every cell's DNA.

At the end of the nine months, cellular differentiation will have produced a miniature human being, complete with a tiny liver, heart and two tiny lungs, and a brain that has the 80 or 90 billion neural cells that he or she will use for the rest of his or her life. Over the following 18 or 20 years, the neurons, along with the myelinated axons [p17] and glial cells [p28] will increase in size. Their number will, however, never increase. If anything, it will diminish with time [p211].

Neural development is usually divided into two processes. In part one, mechanisms independent of sensory activity are already working: it's the assembly phase, ruled by the biological factory (food, sleep, physical exercise and the mother's emotions), and especially by the instructions from the DNA. This is *nature*. In part two, equally crucial, development relies on the activation of sensory mechanisms at the very moment of birth. It's the direct experience of the world – tactile and visual, auditory and sensitive – that adds, modifies or eliminates the synapses that distinguish one human being from any other. In other words, *nurture*.

5.2 START-UP

The moment the brand-new brain comes away from the umbilical cord and embarks on its personal adventure, a sensory tempest is unleashed. An abundance of photons reaches the nerve cells in the retina, which send impulses to the primary visual area of the occipital lobe [p58]. The mother's voice produces sound waves, which, once they have reached the inner ear, are converted into electrochemical signals to dispatch to the auditory cortex located in the temporal lobe [p56]. As a result of information coming from the sensory organs, neurons start multiplying synapses [p18]: upstream, dendrites connect to other nerve cells; downstream, axons [p17] connect to other dendrites. It's a master display of neuroplasticity [p69].

Since plasticity also ensures the learning process in adulthood, it stands to reason that, as years go by, with all the absorption of knowledge, the number of synaptic connections should reach record levels. Right? Wrong.

The brain continues this activity of neurogenesis –

manufacturing new neurons – only for the very first months of life, but it is at around the age of three that it possesses the highest number of synapses. According to some estimates, a three-year-old child has approximately one million billion connections: every neuron is connected to an average of 15,000 others. An adult has about half that. It's a strange decision by evolution, creating superfluity only to begin a process of weeding out redundant connections.

This process is called **synaptic pruning**. Just like gardeners with trees and bushes, the brain has its own system of cutting off unused connections and, at the same time, strengthening those that are in regular use. This monumental reorganisation goes on for many years, at least until the end of adolescence. It involves not only connections but also whole neurons, which, if they don't send or receive information, no longer have any reason to exist and so die.

Neuroplasticity allows the reorganisation of neurons and synapses in case of damage or the loss of a sense, which explains the rich auditory and tactile experience of someone who cannot see. Even in adulthood, structural modifications on a much smaller scale make the learning process possible. But it is immediately after birth that external input is most crucial for shaping the brain of a brand-new human being.

The cerebral supremacy of *Homo sapiens* depends on the size of the brain (it's not the largest in nature but the largest in relation to body weight) and on the prominence of a cerebral cortex, the frontal lobe in particular, dedicated as it is to complex functions such as abstract thought, language, empathy and morality.

No other species on this planet has such a long childhood and adolescence, stages which are clearly necessary for building the architecture of intelligence, consciousness and self-awareness [p129]. The size of the brain, development of the cortex and long childhood are closely connected. From the evolutionary point of view, it's the ability to nurture that has favoured the development of an efficient brain, with a cortex capable of functions that only a long period of construction can render sufficiently complex. The environment and social interactions after birth determine the quality of the human brain machine as much as chromosomes do before birth. Human puppies need much, much more care than the puppies of any other species.

Women and men have been raising children for millennia, but it's only been in recent decades that science has been delving into this process. Another important discovery, in addition to that of synaptic plasticity, concerns the development of brain circuits during specific development phases, called 'critical periods' [p162].

The foundations of brain structures dedicated to vision and hearing are laid two months after birth; those relating to language and speech at around seven months. The basis for more complex cognitive functions, on the other hand, is constructed synaptically at around two years of age. There is a hierarchy in the development of neural pathways because every brain matures in a different sequence and at a different rate. In the case of vision, for instance, the areas that analyse colours, shapes and movement are completed first, before more complex functions, such as recognising a face or interpreting the meaning of its expression. Current science tells us that there are vast opportunities for exploiting the critical periods when such development occurs. As far as risks are concerned at these critical periods, all we need is history to enlighten us.

Frederick II, Emperor of the Holy Roman Empire, spoke six languages fluently. Passionate about science and knowledge, he was particularly curious about man's original language, the one that had been imparted by God to Adam and Eve. He therefore conducted a radical experiment: he had a group of newborn babies raised in isolation, without anyone, not even those who fed them, speaking a single word to them. We can imagine the result: no language, ancestral or other, was ever spoken by those unfortunate subjects.

A complex brain machine like the human one must be started up with much care and attention. For the project written in your genes to be properly expressed, you need a suitable environment and appropriate experiences. This environment includes the absorption of correct nutrients without toxins (during those infamous nine months, and during breastfeeding too), as well as healthy and not overly stressful social surroundings. Sensory experience, which begins gently in the mother's womb, explodes when the baby sees a smile, hears a voice, tastes and smells milk, feels the warmth of a hug, and other sensations throughout the natural unfolding of the development stages. One thing seems clear: the cornerstones of brain architecture are laid even before the child starts school.

The culture of the times influences the way new generations are brought up. At the beginning of the 20th century, long after the lunatic experiments of Frederick II, nobody paid much attention to the brain growth of children, who simply 'grew up'. This reinforced the social gap between those who had the opportunity to stimulate the development of their own brains – through reading, travelling, going to the theatre – and those who did not.

Nowadays, on the other hand, there are those who claim that the periods critical to brain development are over by the age of three. The unfortunate fact remains that the vast majority of education systems don't take the least bit of notice of these – after all – recent discoveries in cognitive science. If they did, then serious teaching of a second language would start at kindergarten and no later.

5.3 ENERGY REQUIREMENTS

A human brain weighs approximately 1.4 kg. This represents about 2% of the total body weight. And yet it accounts for 20-24% of the basal metabolic rate, the energy expenditure of a resting organism. Bearing in mind that this figure varies according to the size of the body, age, sex and state of health, it's still true to say that the brain is hungry for energy.

The brain is your most metabolically active organ. The reason it uses so much energy is because it constantly has to recycle neurotransmitters, as well as to restore the ionic gradients of neurons (the concentration and electric charge of ions moving across their membranes) after they have 'fired' like machine guns at intervals of just a few milliseconds.

If we assume a **basal metabolic rate** of 1,300 kilocalories over the course of a day, that means the brain uses slightly over 56 calories per hour, equivalent to 63 watts; 20% of this works out at 12.6 watts, far less energy than is used by an old incandescent lightbulb. They say that Watson, the IBM supercomputer that, in 2003, beat the human champions of the American quiz show *Jeopardy!*, used 80,000 watts [p242]. It is hard to disagree, then, that the brain is spectacularly energy efficient.

The metabolism of memory and intelligence cells requires nutrients, long periods of rest and short periods of movement. In view of the most recent scientific discoveries, every user should, of course, remember to do this sensibly and intelligently.

5.3.1 Nutrition

There's an extraordinary link between the plant world and our brain world. The sun, the great nuclear reactor at the centre of our planetary system, sends to the earth the energy required to sustain everything every day, from photosynthesis to thinking.

Plants exploit the energy of photons that rain down from the sun in order to reorganise the atoms of six molecules of carbon dioxide collected from the air and twelve from water from the soil. In this way they produce (along with discarded water and oxygen) a glucose molecule that is converted into an energy reserve made of longer chains of carbohydrates and used as food.

The brain, too, feeds on glucose. Through a process that is the opposite of photosynthesis, glucose is produced from the breakdown of carbohydrates and carried by the bloodstream across the blood–brain barrier [p33], where it provides a constant supply of energy to the neurons. Thanks to a chemical reaction that requires oxygen, the glucose is converted into adenosine triphosphate (ATP), the molecule that supplies cells with the chemical energy essential to the body's metabolism.

The energy consumption of the brain – about 120 grams of glucose a day – is essentially stable over the 24-hour period, although the active regions of the cortex burn a little more energy than the inactive ones. (It's thanks to these scarcely perceptible variations that we are able to study brain functions in real time, with PET and fMRi technology [p234].)

The real exception is after a long fast. The brain stores only small reserves of energy and, when glucose is no longer available, it prolongs its function (and survival) by using its back-up supply, an alternative fuel. In such crises, it exploits the energy of so-called ketone bodies, water-soluble molecules synthesised on demand by the liver whenever the glucose reserve light starts flashing. It's not good for it to flash for too long, though: hypoglycemia (the lack of glucose) can cause loss of consciousness and, if prolonged, irreversibly damage the brain.

You would think that the more sugar you eat, the happier your brain. We have the feeling that this is true: no sooner do the sweet receptors of our tongue detect delicious chocolate ice-cream than the brain releases endorphins and dopamine [p25], which infuse their owner with a sense of wellbeing. But this feeling is deceptive.

When you eat a particularly delicious dish, the levels of dopamine that tickle the reward system [p141] soar. If, however, the same dish is served five days in a row, for breakfast, lunch and dinner, then the peaks of dopamine gradually diminish. In other words, the brain encourages its user to be omnivorous and to eat diverse foods that can supply the body with all the macro and micro nutrients it needs. On the other hand, repeatedly eating sweets doesn't alter dopamine levels, and anyone who gets too used to sugar-induced wellbeing can often crave it. This is called addiction [p193].

In actual fact, there are carbohydrates and then there are carbohydrates. Complex carbohydrates, found in natural plant-based foods (and milk) have long chains of sugar, which, during digestion, gradually break down to produce glucose that then enters the circulatory system a little at a time, like a slow-release tablet. Simple carbohydrates, found in refined grains and processed foods, are composed of shorter chains that break down quickly and enter the circuit much faster, like injections. The trouble is, if there is too much glucose about, the pancreas immediately releases insulin, which encourages all the body's cells to store glucose for future use. All except neurons, which, being the only cells without storage space, find themselves with an unstable supply of glucose.

'Our research,' says Fernando Gomez-Pinilla, professor at UCLA, 'shows that our way of thinking is influenced by what we eat.' In an experiment with mice, he found that 'a prolonged fructose-rich diet alters the brain's ability to learn and memorise'. In other words, nutrition habits determine the health and efficiency of the brain. When glucose levels are too low, psychological processes such as self-control [p174] and decision-making [p171] are compromised. However, when they are too high, the entire system slows down. There are other foods that can cloud your judgement too, causing memory lapses and difficulty in concentrating. In lactose-intolerant individuals, for example, eating dairy products can often cause 'brain fog'.

Gomez-Pinilla and his team conducted a further experiment. In addition to huge doses of fructose, some mice were administered omega 3 fatty acids, commonly found in salmon, walnuts and linseed. It was observed that these 'help minimise the damage', countering the effects of too much sugar.

Much has been said in recent years about omega 3; it's almost been considered a miracle cure for all manner of conditions, from cancer to cardiovascular disease, autism to depression. Although we do not have good evidence to prove this when it comes to tumours and heart attacks, there have been encouraging, more or less conclusive studies, about the positive effect of omega 3 on brain function, on cognitive processes in the ageing brain, as well as in young brains with attention deficit [p157] and aggressive behaviour. The findings are more than sufficient to suggest that everybody should regularly eat salmon and other oily fish, like herring, anchovies and mackerel, which contain more EPA and DHA (two different types of fatty acids) and fewer metals (like the dreaded mercury). This is especially recommended for pregnant women.

• For practical advice, see the section on Recommendations [p93].

5.3.2 Sleep

The electric lightbulb radically changed the world. To realise this, all you have to do is visit an African village without electricity and discover, possibly to your surprise, that you can't read after sunset unless you're next to a quivering flame. Thomas Alva Edison's invention has allowed billions of people to spend all 24 hours of the day as they please, having brain experiences at night which their ancestors could only enjoy during daylight hours. There's just one drawback: artificial light has broken sleep.

Sleep is a clear loss of consciousness, a kind of *cogito-er-go-sum* switch: sleep inhibits sensory perception of the world. The brain continues to work, but on a sort of stand-by. Even the voluntary muscles doze off into an experience eulogised by poets and blessed by anyone who drops, exhausted, on the bed, possibly jet-lagged.

However, sleep – evolutionarily so ancient that we share it with mammals, birds, reptiles and fish – has become at times an optional feature, at times an inconvenience – precisely because of the lightbulb. 'Sleep is an insane waste of time and a heritage from our cave days,' Edison proclaimed, proud of his own hyperactivity (and with a real conflict of interest, since he was the founder of the Edison Electric Light Company). One and a half centuries later, politicians and high-ranking administrators publicly boast about sleeping only four or five hours a night. In actual fact, the respective public and shareholders should be worried by this cerebro-muscular exhibition. If deprived of sleep, the brain doesn't work properly.

But why do we fall asleep? A difficult question, simply because there are so many theories circulating. We know for certain that when the brain is deprived of sleep, it doesn't work as well, makes mistakes; its owner is irritable, less creative, and, if the deprivation becomes extreme, it has an extreme consequence: death.

There's the theory that sleep is needed for preserving energy, but this is not very convincing, since few calories are saved during sleep. Some believe that, throughout evolution, sleep has provided a protection and safety function, but this comes up against the fact that we are inert during sleep. There is definitely an element of physical recovery: it has actually been proved that sleep accelerates the healing of wounds and strengthens the immune system.

Yet discoveries in recent years show that, first and foremost, sleep spring-cleans, restructuring long- and short-term memory [p64], thus contributing to the learning process [p162]. According to a recent study, it also literally cleanses the brain of dangerous toxins by means of a kind of hydraulic apparatus known as the **glymphatic system** (so-called because it involves the glial cells [p28] and resembles the lymphatic system), capable of washing away undesirable molecules, such as amyloid beta peptides, which are found in large quantities in Alzheimer's disease. The study found that the less deep sleep a person has, the less his or her brain is able to clean out this harmful toxin, and the more it accumulates.

Thanks to the knowledge gleaned over recent decades, the notion of the apparent uselessness of sleep has been totally debunked. It's not true to say that the brain isn't working during sleep. In some ways, it works even harder. Electroencephalography, the technology that records the electrical activity of the brain, was invented in the 1920s. And yet it was only in the 1950s that scientists used it to confirm that sleep is not a homogeneous phenomenon, but instead has a specific sequence of alternating stages.

You're in bed, reading a book. The waves going through

your brain are probably beta. Then you take off your glasses and switch off the light: the rate slows down, producing the characteristic alpha waves. Gradually, you feel tired and theta waves slow down the rate more and more until they become delta and you're in a deep sleep.

It doesn't end there. In 90-minute cycles, the waves alternate between the absolute immobility of delta and the liveliness of theta, while your eyes are moving at vertiginous speed. This is **REM** (rapid eye movement) sleep, the most peculiar phase of our sleep – if nothing else because this is when we experience our most vivid dreams. The brain's activity is in full swing, almost as though it was watching the screen in a cinema.

When we begin to talk about dreams, things become even more complicated. First, because there is no scientific consensus on what the dream state is, or what it's for. Second, because (since the time of the Ancient Egyptians, who considered them messages from the gods) dreams have been attributed a subconscious function (Sigmund Freud called them 'the main road to the unconscious'), a metaphysical role (contact with the supernatural, such as premonitions) and a paranormal capacity (communication with 'the beyond').

We know for sure that not only dreams, but also dreamers, are totally different from one another. Some people often remember their dreams, others almost never. But soon we all forget almost all of them, especially if they are boring. Most people dream in colour, but some in black and white, like on an old TV set. Individuals blind from birth have dreams with highly developed auditory, tactile and olfactory experiences, and those who have become blind, even when very young, can still dream images. Apparently, negative and anxiety dreams are markedly more frequent than happy ones. And we know that it isn't true, as was believed for a time, that dreams occur exclusively during the REM phase of sleep: it is also possible to dream during the three non-REM phases, although this occurs more seldomly.

It has been proved that REM sleep and dreams play a specific role in readjusting stored memories, removing redundant and unnecessary synapses in order to preserve important pieces of information while deleting useless ones.

In different human cultures, sleep takes on different forms. A perfect example is the *inemuri*, the nap taken by the Japanese in their workplace: partly to refresh themselves and partly to show their superiors that they're working themselves to exhaustion. In any European or American office, this behaviour would be unacceptable, but in Japan it is smiled upon. Just like the legendary Spanish *siesta*, it's been shown that a short sleep during the day is quite beneficial. There are those who claim that there's a scientific basis to the power nap, provided it doesn't last too long (10-20 minutes) and you avoid drifting into deep sleep. They say that Leonardo da Vinci, Isaac Newton and Nikola Tesla slept little and used napping and sleep in general as a resource to fuel their creativity [p167]. Dmitri Mendeleev imagined the structure and layout of the periodic table of elements while asleep in his laboratory chair.

Nobody would ever want to go back to live in a world without electricity, which has multiplied nocturnal neuronal experiences. However, the biology of your brain has not changed: the best thing you can do at night is sleep (and maybe a little during the day as well).

• For practical advice, see the section on Recommendations [p93].

5.3.3 Physical Exercise

Mens sana in corpore sano – everybody knows that. The words written by Juvenal in his *Satires* are so famous that they have become part of everyday language. Because it's obvious that the mind is well when the body is well.

What's odd is that this motto, now indelible in our culture, has been interpreted over time to favour the supremacy of either the *mens*, the *corpore* or the tight connection between the two. As the years have gone by, the last of these seems to have established itself as the best interpretation, not so much due to poetic linguistics as to popular intuition: the mind and the body that carries it around the world are one and the same.

If it's true that constant work (along with rest) is required to build an elastic synaptic structure accustomed to learning and memorising – the perfect image of a *mens sana* – the brain also needs the body to be adequately exercised, so as to obtain a constant flow of biochemical effects.

We know by now that regular aerobic activities (such as running, swimming, rowing, cycling, walking briskly and also dancing) confer a series of benefits to the brain, both in the short and long term.

In the short term, aerobic exercise raises the heart rate (thereby flooding the brain with more oxygen) and helps alleviate stress (which is really not good for the brain) [p196], encouraging the release of molecules that promote psychophysical wellbeing – certainly antidepressants. These molecules look like a catalogue of drugs, albeit legal ones: *beta*-endorphin (an opiate), phenethylamine (a stimulant) and anandamide (a cannabinoid). The last of these, in particular, named after the Sanskrit *ananda*, meaning 'joy' or 'bliss', plays a role in the sense of euphoria you feel if you regularly go for a run, and is also known as 'the runner's high'. No wonder there are a few moderate withdrawal symptoms during abstinence from running.

In the long term, physical exercise enables the mind to give the best of itself. Numerous scientific papers support the idea that regular aerobic exercise makes a positive contribution to operative, spatial and explicit memory, and that it improves attention and cognitive flexibility (known, these days, as 'multitasking').

In response to evolutionary triggers that have shaped the body and mind of *Homo sapiens*, physical exercise contributes to neuroplasticity, as well as the growth and survival of neurons. As long as aerobic exercise is kept up over time, the grey matter in various brain areas increases, especially in the prefrontal cerebral cortex [p51] and the hippocampi [p45], areas associated with the executive functions of the brain.

For once, there is no scientific controversy: the chicken of physical exercise lays the brain wellbeing egg. But what came first, the chicken or the egg?

The theory has been ventured that two million years ago, when our distant ancestors changed their lifestyle to become hunter-gatherers, they needed to carry out a large amount of extra aerobic activity. Not on the treadmills of gyms, but chasing after prey in the forest. In addition to the change in diet, which at that point included much more animal protein than in the past, there is some evidence that physical exercise may have contributed to the substantial growth of brain mass, the cortex in particular, gradually improving cognitive functions.

In any case, no more proof is needed that *mens* and *corpore* are connected inextricably.

• For practical advice, see the section on Recommendations [p93].

5.4 RECOMMENDATIONS

A balanced diet, sleep and physical exercise are the three basic energy requirements for the correct functioning of your brain. But there's more. All three also have the potential to alter your brain health and cognitive abilities.

The attitude to food, sleep and sport varies from country to country. Over the years, science, too, has often changed its mind. But after an amount of trial and error, it is gathering an increasing number of certainties. Here we list a few recommendations that are currently considered valid and, in some cases, merely common sense.

In order to function correctly, your brain needs to know that food, sleep and movement makes exogenous molecules (produced outside the body) and endogenous molecules (produced internally) that are essential for developing, fuelling and supplying energy to the machine that it is.

Nutrition	Sleep	Physical exercise
The standard recommendation is to restock the digestive apparatus with a **varied and balanced diet**, including a lot of fruit and vegetables and very little red meat.	The standard recommendation is to sleep **eight hours a night**, but no fewer than seven. Children and adolescents (if they can manage it) should sleep at least nine hours.	The standard recommendation is **half an hour of moderate aerobic exercise** almost every day. At least two and a half hours a week in total.
The **need for glucose** is met by carbohydrates such as wholemeal bread, and not by sugary snacks that don't provide sugars in a manner appropriate for the functioning of the machine.	If you have problems sleeping, it's good to follow your **circadian rhythm**: avoid bright lights in the evening and, if possible, go to bed at the same time every night.	A brisk walk may be sufficient. But also activities such as running, swimming, rowing, cycling, dancing or heavy gardening are effective.

Nutrition	Sleep	Physical exercise
Great importance has, in recent years, been placed upon **fatty acids**. Omega 3, found in salmon, and other oily fish, walnuts and linseed, is particularly beneficial, especially in the ageing process.	An overly large evening meal, coffee, alcohol and cigarettes **interfere negatively** with sleep. Darkness, silence and a good mattress, on the other hand, can be helpful.	In the long term, aerobic exercise is considered antidepressant and, in the short term, productive of euphoria. It encourages the production of endogenous molecules that resemble a **catalogue of drugs**: stimulants, opioids and cannabinoids.
Drinking water is required because adequate hydration helps the brain to function correctly.	A **nap** of no more than 20 minutes halfway through the day is beneficial to your cognitive system.	Aerobic exercise increases the heart rate and, consequently, the flow of **oxygen** to the brain.
Your brain should know that nutrition has a direct impact on your cognitive abilities, your memory and the correct functioning of the central nervous system in general.	**Your brain should know** that sleep is not optional but required for consolidating memories and cleansing the brain of toxins. It cannot be neglected for long.	**Your brain should know** that physical exercise promotes neuroplasticity, neuron growth, cognitive functions and memory.
Nutrition has an influence on sleep and physical health.	Sleep has an influence on the wellbeing of the body and mind.	Physical exercise has an influence on sleep and mental health.

Other recommendations for excellent functioning:

- Social interactions [p132]
- Limiting stress [p196]
- Meditation [p224]
- Lifelong learning [p216]
- Positive outlook [p129]

6.0 OPERATION

YOUR BRAIN IS MAINLY AUTOMATIC. YOU don't need to remember to breathe or to maintain your heart rate. All the sensory peripherals, such as your skin and ears, are constantly switched on. The flow of emotions is constantly flooding your brain, mostly through no decision of yours, and possibly without you even being aware of it.

Over the years, during the constant interaction with its surrounding environment, the brain forms its own notion of life and the world, and its own unique personality. That's without taking into account everything that occurs below the threshold of consciousness, beyond the boundaries of self-awareness. The **subliminal brain** (from the Latin *sub limen*, 'below the threshold') is what we call the ensemble of unconscious mechanisms that influence cerebral response – thoughts, words and actions – without the user being remotely aware of it. The boundary between voluntary and automatic behaviour is so blurred that some claim – and not without grounds – that **free will**, practically a banner of *sapiens* civilisation, doesn't even exist [p145].

The mosaic of the brain's numerous involuntary and semi-voluntary functions is more than complex. Even if we set aside all philosophical questions regarding our ability to control our destiny, we must still recognise that not all the automatic functions of the brain are totally out of our control.

You can refine your sense of smell, as perfumers do [p97], learn to appreciate music [p106], dominate fear [p116], give the right level of importance to love [p119], develop your own personality [p149], and be more understanding towards your fellow humans [p132]. And much more.

6.1 SENSES

The sensory organs, the largest of which is the skin, are connected to specific areas of the brain, to which they send a frighteningly rich flow of information all at once: sensations. Up in the brain, this information is put in the right order and give it a meaning, turning it into perceptions.

It's inaccurate to say that the only senses are the traditional ones: vision, hearing, taste, smell and touch. We have, in addition, the perception of how our body is positioned in space (**proprioception**), the sense of balance (**equilibrioception**), the sense of pain (**nociception**), the sense of vibration (**mechanoreception**), and the sense of body temperature (**thermoception**). And many more, often inside the body, such as the sense of fullness after a meal or, by contrast, the sense of hunger. The complexity of these systems, often cerebrally overlapping or closely connected, is mind-blowing.

Human beings do not hold all the exclusives on senses, far from it. There are insects that see ultra-violet light, bats and dolphins that use sonar, snakes that are sensitive to the warm blood of their prey, sharks that sense electric fields, birds that navigate using the earth's magnetism. Not to mention the fact that the human sensory and perception system can even be prone to making mistakes [p186].

And there's more. Senses don't convey reality to the brain, but actually *translate* it. There are no colours, sounds or smells in the world. There is the electromagnetic radiation of photons, which the receptor neurons of vision interpret as colours. There are longitudinal waves that compress the air and are transformed into sounds by the auditory neuron receptors. There are fragrant molecules that attach themselves to the olfactory neuron receptors, producing the special effect of perfume.

In addition to this review of those five main senses and their extraordinary properties, let's add to the list – without calling it a 'sixth sense', obviously – the sense of time, or **chrono-ception**. Why? Because it's an essential factor in the perception of the world, recomposed by the brain as a constant flow through a chaos of information that reaches it, every fraction of a second, from the surrounding space. It's the sensation of existing.

6.1.1 Smell

In the beginning was the nose. The sense of smell is the primordial sense, the piece of perceptive equipment with the most distant evolutionary origin. Unicellular life forms developed the ability to 'smell' the pH of the liquid in which they were moving – that is, the varying degrees of acidity on their environment – and subsequent animal species have maintained similar, albeit much more sophisticated equipment, still based on chemistry.

It's true that human beings don't use their noses as much as dogs and mice, but it's also true that many consider the sense of smell almost a secondary sense, which isn't the case at all. The innate ability to measure the environment chemically has contributed to the survival of our species and all those that preceded it. Even today, the olfactory system installed in your brain helps it to get closer to pleasant smells and move away from unpleasant and possibly dangerous ones. Still, being firmly connected to the limbic system [p42], it has the ability to call to mind memories of a distant past, access emotions in the present and, especially, to operate at a subliminal level – without the user being aware of it – for instance, regarding strategic issues such as choosing another human being with whom to undertake the continuation of the species.

The proof that the sensory world starts with the nose is provided by the **olfactory neuron receptors**, one of the most extraordinary variations on the theme of nerve cells [p12].

- The olfactory neurons are equipped with almost 450 different receptors, each with its own keyhole, for which scent molecules – the volatile substances that float above a freshly baked cake or the perfumed neck of a woman – are the keys that unlock the electrochemical message. The fragrance of coffee is produced by almost 1,000 different scent molecules, but the information that reaches the receptors is recomposed by the brain in real time as a single smell. The number of smells distinguishable by a human nose is estimated to be at least 20,000, but some venture much, much higher figures.
- Olfactory neurons are the only ones that literally protrude from the brain. They populate the upper part of your nasal cavity in their millions, sprouting from the **olfactory bulb**, which, located under the frontal cortex, has the task of triaging the information from that multitude of sensors.
- Olfactory neurons are among the very few capable of neurogenesis. As a rule, neurons are born and die with you. However, while the brain is well wrapped and protected by the meninges and the cerebrospinal fluid [p33], the neurons that extend to the attic of the nasal cavity are

exposed to the environment and so are at risk of degenerating. The evolutionary solution is blatantly obvious: make them able to regenerate.

- It doesn't end there. A human genome contains approximately 25,000 genes, i.e. 25,000 instructions for building the complexity of a functioning and thinking body. Well, at least 858 of these are devoted to the 'construction' of olfactory neuron receptors, that is 3.5% of the entire chromosome heritage. Of these, 468 are pseudogenes, which means they are former, now deactivated functions: a mutation has removed their ability to codify a protein. This explains why in humankind the sense of smell, although it has strategic importance, is less sensitive than in other mammals.
- Finally, as definitive confirmation of its distant ancestral origins, smell is the only sensory system in which information does not pass through the thalami [p43]. In other words, it skips over these essential hubs of communication with the outside world ('essential' because interrupting them can cause a coma). Not only is it the only sense that remains switched on during sleep, but it has been proven to work even in cases of total unconsciousness.

It's no wonder the olfactory bulb is considered an integral part of the limbic system: the receptive centre of the nose is well connected to the amygdalae and the hippocampi, i.e with emotions and memory. Marcel Proust wrote a literary monument about the subtle, sweet flavour of *madeleines* that rekindled childhood memories (flavour is largely dependent on the sense of smell [p101]). And yet, incredible as it sounds, humans find it difficult to describe smells with words. Except for the sommelier, who uses obscure, esoteric terms ('austere', 'short', 'decadent') to describe wine, everybody else finds it a challenge to go beyond 'good' and 'bad' and a few other generic adjectives. There are different theories as to why that is, from the quality and strength of the connection between the olfactory system and the language module, to the structural inadequacy of many languages.

On the other hand, there's another direct connection for which there is most definitely no shortage of words. It's the link between the olfactory bulb and the automatic module of sexual reproduction. In 1959, scientists discovered that many animals and insects are able to communicate by means of chemistry through – mostly odourless – molecules called **pheromones**. Vertebrates and many mammals even have a second 'nose', called the vomeronasal organ, which specialises in perceiving signals coming from pheromones. The messages encoded in these pheromone chemical signals are numerous and differ according to species, but mainly revolve around just a few topics: sex, food, opportunity and danger.

And what about men and women? Apart from the fact that recent analysis seems to imply olfactory dimorphism, that is, a structural difference between the male and female olfactory system, despite a plethora of studies, scientists have yet to discover the existence of a human pheromone. In some cases, they have confirmed the presence of evolutionary vestiges of a vomeronasal organ, but this doesn't appear to function. This suggests that the little bottles of human pheromones sold on Amazon at prices ranging from $8 to $150 could be a cute bit of swindling.

While we're on the subject, doubt has also recently been cast on a 1970s study describing the so-called Wellesley Effect (named after the women's college in Massachusetts), the alleged synchronisation of menstrual cycles of women who live at close quarters, regulated by the sense of smell. Still, this does not mean that smell is of no importance in the lineage of the human species. Body smell – determined by genetics, the environment, personal hygiene and diet – plays a large part in the reciprocal decision to mate.

It all begins with the nose.

6.1.2 Taste

The world's restaurant owners, starting with those with three Michelin stars, owe a debt of gratitude to evolution. If your brain is capable of gushing over a bowl of *tortellini alla bolognese* or even simply telling a Cabernet from a Merlot, it is not because our very distant ancestors' appetites had to be whetted before they would eat: hunger already did that for them.

The sense of taste evolved for a much more practical reason: to distinguish edible foods from non-edible ones that might be poisonous or off. We feel that a fried egg would go beautifully with a dusting of truffle, simply because of the prosaic matter of survival.

The tongue is populated by thousands of papillae. Each one (except for so-called foliate papillae) contains hundreds of taste buds. Every one of these has between 50 and 100 taste receptors. Forget the old-fashioned, deeply rooted notion that the tongue is divided into zones specialising in sweet or savoury. That's nonsense: the receptors capable of perceiving sweet, bitter, sour, salty and *umami** tastes are distributed over the entire surface of the tongue. But these five flavours are just a tiny portion of the huge sensory palette called taste.

The sense of taste alone is not sufficient for sensing flavour.

---

*   The concept of *umami*, a Japanese word meaning savoury taste is much more
      familiar to brains that live in the Far East, where the cuisine frequently uses
      monosodium glutamate, also known as 'flavour enhancer'.

Flavour is produced by the sum of information coming from the tongue's receptors and from the more sophisticated ones in the nose, which is processed, in real time, as usual, by your brain. It's not by chance that a bad cold has the power to stand in the way of culinary delight.

But there is more. The brain owes what it perceives while chewing through a bowl of *penne all'arrabbiata* to information coming from the taste and olfactory systems, but also from mechanoreceptors that distinguish the texture of the food, from thermoreceptors that register its temperature and from the mucous membrane that senses the sharp presence of a chilli pepper. Everything is recomposed in the brain as a single experience. Here, too, the power of the neural network is manifest in all its useful complexity. Useful for the survival of the species – and for its well-loaded forks.

6.1.3 Vision

In this precise *moment*, from the billions of billions of photons frantically bouncing around, only a few thousand have leapt from this white page all the way into your retina. Over 100 million photoreceptors – located, strangely, at the back and not the front of the eye – have converted the luminous signals into electrical signals, translating the chromatic language of light into a language that is more comprehensible to the brain, somewhat like the sensor of a digital camera. In actual fact, only a minimal part of the retina, the **fovea centralis**, is able to focus on the word 'moment' by means of a small number of highly specialised photoreceptors that frame the letters of the alphabet in a series of rapid but imperceptible eye movements.

The electrical impulses are carried through the **optic chiasm**, a kind of railway junction, to opposite sides of the brain: information from the left eye goes to the right hemisphere, and vice-versa. Then, having passed through the thalamus, they eventually reach the occipital lobes, 'switching on' the area called the **primary visual cortex**. There are other areas that supply necessary information – often low-resolution or blurred – to the brain, so that the perception of a complete image can be processed, while on the move and in real time. What is incredible is that barely 40 milliseconds elapse between the written word on the page and the visual cortex: that's one 25th of a second.

Charles Darwin considered the evolution of the eye through natural selection to be so astonishing as to be 'absurd in the highest possible degree'. The wonder of seeing is achieved through a succession of optical mechanisms, starting with the cornea – an actual lens that sends the retina an upside-down image – and finishing with the mental mechanisms that involve most of the cortex. The light that hits two two-dimensional retinae projects a three-dimensional panorama in your brain. Isn't it incredible?

Each human retina has about six million cone receptors and about 120 million rod receptors, which actually *translate* the light into electrical impulses. Only cones populate the central fovea, the high-resolution visual area. Along with a myriad of rods, they also populate the rest of the retina. The cones distinguish colour and require many photons in order to be activated, while the rods, essentially achromatic, need just a handful of photons: this is why in a semi-dark environment colour perception is reduced or disappears all together.

Cone receptors produce the magic of colour with a trick replicated by television, which, by means of RGB (red blue green) technology, reconstructs millions of colours by combining the frequencies of three different pixels. Even without specifically using the red, green and blue of the monitor, the brain reconstructs colours through three types of cone receptor, each specialising in picking up different frequencies of the visible electromagnetic spectrum. This phenomenon, which allows you to enjoy a sunset or a painting by Van Gogh, is even more wonderful if you think that the colours themselves are manufactured by the brain. Not only is beauty 'in the eye of the beholder', as the saying goes, but so is colour.

Visible light, then, is produced by photons at a frequency of between 430 and 750 terahertz (THz), oscillating – depending on the colour – between 430 and 750 *million million* times per second. When sunlight illuminates a tomato, the skin's chemical components are able to absorb a large proportion of the radiation, but not frequencies around 500 THz, which are therefore reflected. In your retina, cone receptors that contain a protein called **opsin**, capable of responding to this frequency, are activated, and these produce the perception of red in the brain. A courgette reflects frequencies of around 550 THz and a blueberry of around 650, activating the two other kinds of receptors, each equipped with their own respective opsins. The 'yellowness' of a lemon is generated in your brain by a frequency between green and red, activating both receptors in different measures. A glass of milk, in its whiteness, activates them all in equal measures.

If, by any chance, you are colour blind (a genetic defect whose most common manifestation is confusion of the colours on the green-yellow-red spectrum), it's because you do not have one of these three receptors. On the other hand, many animals, mostly birds and insects, are capable of perceiving ultra-violet light, because they have a fourth kind of receptor, capable of capturing those photons that oscillate even more rapidly than violet. In their eyes, the colours of any flower are *completely* different from the ones you see.

For years, there was the belief that the human visual system received a series of images in sequence, a bit like a movie camera, which can record 24 or more images per second and so, in the cinema, give the impression of a constant flow (that's right, the Hollywood empire is built on an optical illusion). Later, however, it was discovered that the brain has its own strategy for rationalising the enormous flow of data that arrives every millisecond from the sensory organs: the visual cortex cuts excess information and saves energy by transmitting only the differences in the image. It's more or less what algorithms do when compressing video data, when they manage to reduce the 'weight' of bits – the atoms of the digital world – and transmit them along the arteries of the Internet. The visual system is obliged to use the available resources as best it can, because of an inherent problem: the photons hitting your retinae as we speak are carrying much more information than actually reaches your occipital lobes.

That's right: the mind-blowing wonder of your visual system is full of defects, incongruity and redundancies that illustrate the convoluted path of evolution. Forget the well-known defects, such us short-sightedness and astigmatism. The fovea centralis is tiny, takes up two degrees of the visual scene at most, and the brain has to solve the problem with saccades – jerky and frequent eye movements – in order to frame what it wants to see. On the retina, at the spot where the optic nerve begins, there are no receptors, so there's a **blind spot** where the eye doesn't work, and yet the brain reconstructs a blurred and approximate image at a subliminal level, so that you don't get to see the two black holes that would otherwise be at the sides of your field of vision. If we remember that 3D vision is an optical illusion and that colours are subjective rather than objective, the eye does begin to seem, as Darwin said, 'absurd in the highest possible degree'.

After travelling through the hierarchy of visual cortexes that recognise first the outlines of the image, then its colours, then its movement and position in space, 'neurovisual' information reaches the parietal lobes to compute spatial data and the temporal lobes to recognise objects, and especially, recurring patterns. Your brain has a true fixation with pattern recognition. This application, which was installed millions of years ago, has become essential to a function that's indispensable to social interactions: face recognition.

Your brain sees faces everywhere. In the clouds, the moon, on dirty walls and in puddles. The automatic application scans every face it passes in the street for just a few milliseconds, then identifies it as familiar/unfamiliar, resembling/not resembling, woman/man, attractive/ugly and much more. This pattern recognition doesn't just involve vision; in some rare cases, errors in a range of perceptions can lead to identifying patterns that don't actually exist. This is called **apophenia** [p190]. Some people find patterns in the numbers of the lottery draw, others in a stain on a wall that vaguely resembles a religious figure, others in the divinations of a tarot reader. According to some researchers, including Michael Shermer, author of the book *Homo credens*, this automatic application is one of the many that contribute to the development of beliefs or faith in highly improbable things.

'I won't believe it till I see it,' you hear people say. Except that the brain also believes what it does not see. And what it does see is mostly a magnificent illusion.

6.1.4 Hearing

What sound is that? Where is it coming from? The most distant evolutionary origin of hearing consisted in finding an answer, preferably immediate, to these two questions. Two questions linked to survival. Is there danger nearby? Where exactly?

The sense of hearing, which originated as early as the first amphibians millions and millions of years ago, has long been used for intercepting prey as well as averting the danger of becoming it. Based on both grandiose and miniature architecture (the modern ear is made up of dozens of components, with thousands of parts working in sync), the ability to hear proved to be critical to developing that refined property that led to the evolution of *Homo sapiens sapiens*: language. And perhaps, even earlier than that, to incubating the most mysterious of its unique brain abilities: music and the pleasure of listening to it.

Sound is a wave that travels through the air. If, in a science fiction film set in the void of space, the director plays you noises, be sure that he is taking you for a ride: if there is no air, there is no sound. The sound wave that makes the ether vibrate travels at approximately 1,230 kilometres per hour and causes the **hair cells**, the sensory neurons of the hearing system, to vibrate. These are located inside the organ of Corti (named after the Italian anatomist who discovered it in the nineteenth century), a section of the inner ear, and lie against the **basilar membrane**, which is able to resonate like the strings of a musical instrument. As a matter of fact, the whole mechanism looks like a musical instrument.

When playing the solemn A at 440 hertz, as orchestras do to tune, a specific area vibrates 440 times per second. If we play the A an octave lower, in other words at 220 hertz, another area, further down, vibrates more slowly. Using this information, the brain's auditory cortex – located in the two temporal lobes, just above the ears – reconstructs the frequency, speed, intensity and even direction from which the sound is coming. In other words, anything from the roar of a lion behind a bush to a love song by Frank Sinatra.

Among the many mysteries of neuroscience, there's music. Why does a beautiful song trigger the release of dopamine [p25], making human beings take pleasure in listening to it? Why does a string quartet contribute to lowering the levels of cortisol [p27], the stress hormone, and increasing those of immunoglobulins, antibodies crucial to the functioning of the immune system? After all, there doesn't seem to be a close evolutionary link between music and natural selection.

It has long been believed that the brain's pleasure in music is based in multiple neural areas reserved for totally different functions. In 2015, researchers at MIT located an area in the auditory cortex that responds specifically to music and not to other sounds. Another piece of research, led by the University of Jyväskylä, in Finland, noticed − also using fMRI technology − that music lights up the brain significantly beyond the temporal lobes [p56]. Rhythm, one of the essential components of music, involves the motor areas of the brain, revealing a strong connection between music and dance. Melody, the succession of frequencies at precise mathematical, tonal and temporal intervals, involves the limbic system and therefore the emotional centre. Harmony seems to be associated with the default mode network, a series of cerebral areas active during the apparent rest phase, on which the mind's ability to wander, and creativity in general, seem to depend [p167].

Music is a universal neuro-experience, in the sense that it involves all human cultures without distinction. But science has also proved that making music is beneficial. It seems that in musicians, the corpus callosum and the areas devoted to motor control, hearing and spatial coordination are more developed than in non-musicians. They say that making music collectively, synchronising the tempo with other musicians and singers, triggers the release of oxytocin, the so-called attachment hormone [p26]. Could this be because our distant ancestors partook in ceremonial singing before a battle or a hunt, and the group sense encouraged by oxytocin gave them a competitive advantage? Even nowadays, those who regularly sing in a choir claim that singing with other people has a positive effect on their health.

On the subject of competitive advantage, human hearing and the cortex that decodes it have produced another: the understanding of languages, probably the most important characteristic that distinguishes *Homo sapiens* from other animal species. Not just Wernicke's area (specialised in language comprehension) and Broca's area (language production) [p51], but many other neural pathways are involved in converting sound waves into words full of meaning, associations and categories. Words which, when placed in a specific sequence, are even able to stimulate a flurry – or even a storm – of emotions in the limbic system.

Few words are needed to praise the impact that instant communication has had on human evolution and its civilisations. If there was no hearing, we would have had to invent it so we could talk.

6.1.5 Touch

It always comes at the bottom of every list of the five principal senses. We call it touch, but that's a word that cramps its style. The **somatosensory system** – a more pompous and appropriate name – transmits a large amount of very diverse information to the brain, which comes from every part of the body – a body literally strewn with sensors specialised in different tasks. As well as for touch, there are receptors for pressure, pain, temperature, vibration and balance, which send signals from the skin, muscles, bones, internal organs and also the cardiovascular system. And now try and guess: which is the only part of the body that does not have these sensory properties? Which is the organ, which, if pierced with a pin, does not send any pain signal to the brain? That's right, you've guessed it: it's the brain itself, which doesn't even have one single pain receptor.*

It's a beautifully sophisticated mechanism. It's an integral part of the peripheral nervous system, where the **sensory** (or afferent) **neuron** [p12] is at work. It has the soma in the spinal cord and an axon [p17] that connects its specialised receptor (let's say for pain), and is able to transform a pin prick to the knee into a signal encoded in the electrochemical language of neurons. The signal – thanks to two other neurons that pass it like a baton on an athletics track – travels through the spinal cord at high speed, reaches the medulla oblongata [p39], then the thalamus triage centre [p43] and finally the **somatosensory cortex** of the two opposite parietal lobes for final processing (mixed with millions of other signals).

The primary somatosensory cortex, divided into four segments dedicated to various operations, receives signals according to a precise map of the human body, in reverse, as usual: those of the right foot converge on a specific point in the left parietal lobe, and those of the left hand on a specific point of the right lobe. Since receptors are concentrated in the most sensitive parts of the body, for example the fingertips, lips and tongue, the space these parts occupy in the primary somatosensory cortex is disproportionate to their real size.

This discovery is almost a century old. In the 1920s, Canadian neurosurgeon Wilder Penfield operated on a few

---

* And what about headaches? – you may well ask. There are other structures in the skull that contain pain receptors, such as cranial nerves and meninges, the membranes that wrap the brain and hold on to cerebrospinal fluid [p32].

hundred people's brains, opening their skulls with just local anaesthetic. Since a pin causes no pain to the brain, he decided to seize this opportunity for scientific purposes. So he began to insert electrodes in the grey matter of his infinitely patient patients in order to judge the effect. This is how he discovered not only that the electrical stimulation of a temporal lobe produces a recollection of past memories, but also that the parietal lobe encodes this detailed, yet disproportionate map of the body.

Picture a humanoid form with enormous hands and feet, as well as a huge tongue and lips: it is the brain's notion of its body. It has been called the *homunculus* and, by doing a simple search on the Internet, you can even see how it was represented in two dimensions by one of Penfield's assistants. There is also a three-dimensional model, where you can see that the space reserved for the genital organ is rather small despite its being, well, a rather sensitive area. Penfield (who refrained from drawing a female *homuncula*) must have been the victim of the prudishness of his times.

6.1.6 Time

Unlike all other machines, including the microwave, the brain does not have a clock that beats the inexorable passing of seconds. But this doesn't mean time – the fourth dimension – has no importance in the darkness of the cranial box. On the contrary.

Time is an essential component of the sensory system, because it supports the continuum of experiences that is at the root of self-awareness: every user clearly feels that he or she is always the same person, be it an hour ago, now or an hour from now. This is why, even though the brain does not have an organ dedicated to recording the seconds that elapse, we decided to include time among the principal senses: it gives human life meaning.

The brain has a series of components for detecting time, which – unlike a computer clock – is not absolute but relative. In other words, it depends on the person who experiences it. It is believed that the perception of time is governed by a distributed system that includes the cortex [p51], the cerebellum [p41] and the striatum [p48], as well as by the information that arrives non-stop from the five senses.

Even though the jury is still out, the fact that, when we are young, the days, months and years seem to flow slowly, and then acquire increasingly vertiginous speed when we reach adulthood, is usually explained by the different amount of information that floods the brain. To a child's brain, the continuous sensory experiences are always new, and so create continuous neuroplastic configurations [p69]. To an adult's brain, on the other hand, they are mostly repetitive and synaptically less noteworthy. This peculiar difference gives you a different estimation of the time you still have available, and often turns out to be a terrible hoax: when you're young, time flows slowly and appears to promise a very long life; when you grow up, you realise that the days and weeks are whizzing by.

This is just one of the countless illusions in your perception of time. You know very well from personal experience that time seems to fly when you're doing something interesting or enjoyable. On the other hand, when you're bored or uninterested, minutes drag at a snail's pace.

A large number of psychology experiments have shown a certain regularity in the imprecise perception of time: people tend to remember recent events as though they were more distant, but also to estimate distant events as though they were more recent than they actually are [p189]. And during a road accident or any dangerous situation, time seems to slow down.

It was personal experience that led neuroscientist David Eagleman (who fell off a roof at the age of eight) to study this phenomenon in depth and reach the conclusion that the slowing-down-of-time effect is just the feeling linked to the memory of the accident, because in such circumstances memories are 'packed more tightly' and the action is relived as though in slow motion. In any case, the fact that time seems to expand in situations of danger is an illusion with obvious evolutionary implications, since our distant ancestors used to risk their lives as frequently as we clean our teeth.

The perception of time is influenced by the age of the brain, the surrounding circumstances, the neurotransmitters in action and a wide range of psychological factors. And also by sunlight. In actual fact, the brain does have a kind of clock. It doesn't keep track of minutes or even hours, but can record days within the flow of dawns and sunsets. This clock, called the **circadian rhythm** – 'circadian' means 'about a day' – is located in a nucleus of hypothalamic neurons [p46] that regulate the constant flow of changes in the brain through an arc of 24 hours. Light is the main switch in this system, capable of turning on and off the genes that set the entire organism's rhythm.

By regulating sleep and hormone production (as well as body temperature), the circadian rhythm has a determining role in the correct functioning of the brain machine. One and a half centuries after the invention of the lightbulb, which has had a detrimental effect on our sleep mechanism [p87], human beings tend to sleep far less than biology intended. The faulty functioning of the circadian rhythm is linked to depression and other illnesses. You just have to experience jet-lag after an intercontinental flight to get an idea of what happens when the time zone is shifted without the hypothalamus being informed...

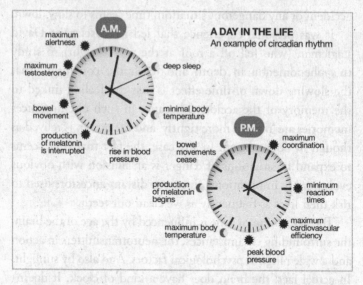

A.M.

maximum alertness

maximum testosterone

deep sleep

bowel movement

minimal body temperature

production of melatonin is interrupted

rise in blood pressure

P.M.

bowel movements cease

maximum coordination

production of melatonin begins

minimum reaction times

maximum body temperature

maximum cardiovascular efficiency

peak blood pressure

In the case of time, as with the other senses, there is a rainbow of individual differences. Every brain belongs to a specific chronotype, or inclination to sleep at different times. At the two extremes, there are those who go to sleep at sunset, and those who do so near dawn.

Time adds a fourth dimension to human life, made up of an amalgam of the past to remember, the present to live and the future to plan. It's not true that time is money, as someone said. Time is the life we have left to live. And even though it doesn't have a clock, the brain knows this.

6.2 FEELINGS, EMOTIONS

Abandonment, admiration, affection, anger, anxiety, boredom, compassion, despair, devotion. Dignity, disagreement, enthusiasm, envy, esteem, exasperation, fear. Forgiveness, friendship, frustration, gratitude, gratification, grief, guilt, happiness, hatred, honour. Indifference, indignation, irritation, jealousy, loneliness, love, melancholy, misandry. Misanthropy, misogyny, mistrust, nostalgia, pity, pride, regret, remorse, resentment. Revenge, sadness, shame, spite, solidarity, trust, wonder...

Here are fifty but there are many more. One reason it's difficult to draw up a complete list of feelings and emotions is because they often overlap, even in terms of langauge. Moreover, their definitions and nuances change considerably from one culture to another. In Danish, *hygge* refers to the pleasant feeling of cosiness when you are enveloped in the warmth of your home, with friends, in front of a fire. The German *Schadenfreude* connotes a totally different kind of pleasure, linked to the misfortunes of someone else. In Mexico, on the other hand, the *pena ajena* is the embarrassment felt when witnessing somebody else's humiliation.

These are words which simply don't exist either in English or in many other languages. Yet one thing is certain: *hygge, Schadenfreude* and *pena ajena*, just like love and hatred, exist only in the brain of the person who experiences them. There's an important additional detail: these feelings are often, if not almost always, triggered by events outside the brain, which are wholly independent of it.

This partial, albeit substantial, inability to control such emotions can be explained by the way in which evolution has organised its connections: there are far more neural pathways going from the limbic system (emotions) to the cortex (rationality) than the other way around.

In order to get a general idea of how they work, let's analyse just three feelings that are nonetheless representative of the brain's function as well as of evolution: fear (an emotion with very distant origins which is, to say the least, crucial for survival), love (essential for reproduction and the care of descendants in mammals) and happiness (which makes the human world go round).

6.2.1 Fear

Fear was invented a few million years ago for a very specific reason: to ensure survival in case of danger. It's an integral part of brain functioning and, in order to guarantee a reliable service, works on automatic and at high speed.

Let's take an example. You're walking in a wood and experiencing considerable pleasure from the information reaching your senses: the song of the wind and the birds, the colour of leaves illuminated by the sun, the scent of fresh air in the underwood. Then your retina is imprinted with a sinister, elongated form, lying on the ground. The information reaches the thalamus [p43], which immediately dispatches it to the amygdalae [p44], the control centre for fear. Even though the data received is still rather vague, the amygdalae order the brainstem instantly to block all the body's movements (to avoid getting closer to the danger), command the facial muscles to open the mouth and the eyes wide (to warn others of the danger) and the hypothalamus [p46] to order the production of adrenaline, which increases the heart rate, blood pressure and breathing (the so-called 'fight or flight' response). What is so great about all this is that the entire process takes about 400 milliseconds, less than half a second: in other words, it happens long before your brain realises that there is a venomous snake two metres away.

According to psychologists, fear is one of the few inborn emotions and it has been preserved throughout the whole of evolution because of its obvious practical result: the survival of the individual and the species. However, as we can see, fear is not triggered by danger per se but by the *forecast* of the danger [p60].

Let's go back to that wood for a moment. While dispatching high-speed information to the amygdalae, the thalamus also sends it to the visual cortex, which processes it and sends it back more slowly to the amygdalae. It was a false alarm: it wasn't a viper but a curved branch that looked like a snake. Its colour and shape alone were enough to trigger the alarm. About a second later, the amygdalae signal the end of danger and everything, heartbeat included, quickly returns to normal (whereas if the presence of the snake had been confirmed, then it would have strengthened the chemical 'fight or flight' signals).

It doesn't end there. The amygdalae also inform the hippocampus [p45] of what's just happened, as well as the prefrontal cortex [p53], which deals with the cognitive and learning process, so as to form a memory that will come in useful in other situations of danger, be it the ancestral fear of snakes or, more probably, of a driver who doesn't respect a pedestrian crossing.

It's the cortex that processes the distinction between the rational fear of a snake and the irrational fear of a dry branch. If irrational fear is chronic, it has a specific name: **phobia**. There are endless lists on the Internet of all the phobias in the world, coined etymologically from the Ancient Greek. Claustrophobia (fear of enclosed spaces), glossophobia (fear of speaking in public) and arachnophobia (fear of spiders) are well known. But there is also ablutophobia (fear of washing oneself), sanguivoriphobia (fear of vampires), nosocome-phobia (fear of hospitals), xylophobia (fear of woods, with or without slithering reptiles), and many more [p199].

Fear is a built-in survival mechanism that has certainly contributed to keeping alive the seven billion plus human beings who currently inhabit the planet. Fear has probably also saved you. If nothing else, because memories of fear always caution us to be careful, for example when we cross the road.

In his book, *The Gift of Fear*, Gavin de Becker writes that 'precautions are constructive, whereas a state of fear is destructive'. It is destructive because 'it can also lead to panic, and panic itself is usually more dangerous than the outcome we dread. Rock climbers and long-distance ocean swimmers will tell you it isn't the mountain or the water that kills – it is panic'. Fear is the expectation of danger, but panic, the fear that goes round and round, reinforces that expectation. It may turn out to be useful to know this. When you're swimming, for instance – without crossing the English Channel or the Straits of Messina – in any choppy sea.

A protracted state of fear can also be destructive in other ways. Stress (a concept and word that originated around 1930 and, as a matter of fact, is called that in just about every language in the world) is closely connected to the mechanism of fear but has a milder effect than the momentary terror at the sight of a snake. Even so, the prolonged stimulation of the 'fight or flight' response causes an excessive presence of cortisol [p27], capable of negatively affecting health and the immune system.

Fear is a standard fixture in every brain – except for those rare individuals who have damaged or atrophied amygdalae and are totally incapable of experiencing the arc of emotions that goes from slight fear to terror. For some, the excessive state of fear may require the intervention of pharmaceutical or psychological treatment. For many, fear can produce desired chemical effects, as we can see in those who love horror films or roller-coasters. In all brains, however, it's very useful to practise distinguishing what is truly dangerous from what is not. Because too much fear is harmful [p196].

As Franklin Delano Roosevelt said – for totally different reasons – in 1933, in the midst of the Great Depression, 'the only thing we have to fear is fear itself'.

6.2.2 Love

'Darling, I love you from the depths of my hypothalamus,' the female biologist said. But the man, who had studied law, took offence and never saw her again.

This could be the perfect micro-novel for the Twitter era, but there's undeniable truth between the lines of silliness: although love is symbolised by the heart, that's not where it resides. Love lives entirely in the brain, with a specific preference for the limbic system. This doesn't mean that every proverb involving the heart must be corrected, but the time has finally come for crude reality to be accepted: love, too, is an entirely neuronal experience.

Some kill for love. Some literally die for love. The experience of falling in love produces on the brain stimuli that are, in intensity, up there with hunger and thirst. This is all thanks to a powerful cocktail of chemical substances that flood the appropriate parts of the brain.

It starts, inevitably, with sexual attraction. Testosterone and oestrogen, that is, male and female hormones, are busy encouraging – or discouraging – dates. But there is also a whole complex structure at work: that of opioid receptors, that network of 'keyholes' distributed in many areas of the brain that allow the body's endogenous opiates to unlock them and, therefore, to function.

In a group of male brains which were administered substances that impair opioid receptors, looking at the photos of beautiful women triggered a much milder neuronal reaction than in normal brains. In other words, for an encounter to flourish, it must pass the hormone and endorphin [p20] as well as the olfactory test, smell being the primordial sense that's deeply involved in sexuality [p97]. According to data collected by scientist Helen Fisher, of Rutgers University, formerly consultant for the dating site Match.com, at least one third of close encounters of the first kind evolve into a love story, i.e. move up to the next level.

The team managed by Fisher, who 'looked' into a series of brains in love by means of fMRI technology, found that the ventral tegmental area [p41] – the dopamine ammunition store – is especially active during the passion phases. Dopamine [p25] produces that sense of desire and excitement typical of 'level two'. But there is also the 'fixation' effect.

Cortisol [p27] – the stress hormone – levels rise to tackle the novelty of a burgeoning love, triggering anxiety and a drop in serotonin [p24]. The lack of serotonin is also associated with obsessive-compulsive disorders [p204], which might explain why the brain in love can't think of anything except the beloved brain, to the point of overlooking the appearance of the first little flaws. Otherwise, there's not much we can do: when the visual systems of two brains converge – in other words, when two people in love look into each other's eyes – adrenaline and noradrenaline [p24] are there, ready to speed up the heartbeat and supply that intoxication that's not too dissimilar from a generous sniff of cocaine.

'Level three', that of a love that persists through time, is regulated by vasopressin, which contributes to long-term rela-tionships, and oxytocin, a hormone and neurotransmitter with an important role in human history: it's the attachment molecule [p26]. So-called romantic love is, as a matter of fact, no more than a happy invention of *sapiens* evolution, whose objective is the reproduction of the species.

From the evolutionary point of view, the stable presence of a father as well as a mother is highly desirable for the healthy growth of a brand-new human being. The total absence of someone to feed, protect and care for him or her is notoriously fatal. To develop such a sophisticated control centre inside his or her skull, *Homo sapiens* requires a long childhood and a long adolescence [p79].

That is why it's not just the orgasm that generates oxytocin. Mothers also produce and administer oxytocin during breast-feeding, thereby strengthening the bond with the new arrival. Even dogs get a discharge of oxytocin when they look into their owner's eyes. It's the extraordinary chemical and psychological power of attachment.

It becomes clear that the most overwhelming feeling in the world, respected and celebrated by art and literature in every human culture in history, is produced by a brain network that is so strategically important for the ultimate objectives of natural selection that it can take over the entire central nervous system – including the frontal cortex, the head office of rationality. Is it a coincidence that the proverb 'love is blind' is the same in at least fifteen languages?

Thanks to its extraordinary pharmacological arsenal, love has the firepower to throw the reward system into turmoil to the point of triggering something akin to drug addiction. The dose of neurotransmitters released by the sight of the person we love can come to feel like a compelling necessity to us, and cause actual withdrawal symptoms in case of a traumatic break-up.

Over time, cortisol and serotonin return to normal levels, the stress eases and 'obsessive' thoughts die down, usually in the space of a couple of years. Dopamine, on the other hand, can carry on dispensing its happy neural reward, and oxytocin levels can stay elevated for decades – as long as we remember how to produce them [pp25, 26].

Such a brief, crude description of love, which runs counter to millennia of songs, poems, elegies, paintings, novels and cinematic masterpieces devoted to the praise of the most beautiful and effervescent feeling there is, may have caused a slightly unpleasant feeling in your brain – for which we apologise. Please note that this was itself an indirect experiment. Since the concept of romantic love is a social invention that varies from one culture to another, and the emotions it triggers can also vary from one brain to another,* this huge amount of artistic production in its honour cloaks it in an aura of sacredness. That's why your brain may have just experienced a synaptic conflict between GABA and glutamate [p23]: as one is inhibitory and the other excitatory, in combination they trigger that feeling of unease.

To explain this feeling better, let's turn to the irritated response of physicist Richard Feynman – one of the best brains of the 20th century – to a friend who was contrasting the artistic beauty of a flower with its stark dissection by scientists: 'All the answers provided by scientific knowledge add something to emotion, the mystery and the wonder of a flower. It's an addition, not a subtraction.'

Love is a wonderfully neural thing.

6.2.3 Happiness

As you may be aware, 8 March is International Women's Day. You are also probably aware of some of the large number of little flags that feature in the United Nations calendar to commemorate worthy goals – Democracy Day, Peace Day

---

* The emotions triggered by love, and even the basic concept of love itself, can vary markedly, depending on the brain model used [p178].

and Nuclear Disarmament Day. The twentieth of March will likely escape your notice. But this date, which is known in the northern hemisphere as the last day of winter, is, 'in order to acknowledge its importance in the lives of people all over the world', dedicated to happiness.

The pursuit of happiness is the most powerful motivator in the world, irrespective of social, geographical and geopolitical differences. It is clearly printed in the 1776 United States Declaration of Independence, although it is not mentioned in the 1948 Universal Declaration of Human Rights. In 1972, after the King of Bhutan started a government campaign to improve Gross National Happiness instead of Gross National Product, the notion of measuring the average happiness index of a nation soon reached the shores of scientific research, politics and international law. Happiness was defined as a 'fundamental human right' in the 2012 UN resolution that created the International Day of Happiness.

Happiness is the most sought-after of all mental states, although it varies according to the semantic definitions – contentment, fun, satisfaction, gratification, joy, euphoria and so forth. Like every other mental state, it depends on a combination of chemical factors (neurotransmitters), electrical factors (brainwaves and action potentials), as well as architectural factors (the structural connections of each individual brain).

On the subject of the structure, it has been observed that the left prefrontal cortex [p53] is particularly attentive to feelings of happiness and that the right one seems linked to sadness. Besides dopamine and oxytocin, which manage the entire reward system [p141] and attachment, a large number of molecules are involved in modulating happiness, from simple good mood to ecstasy. There are endocannabinoids, such as anandamide [p91], molecules similar to cannabis but manufactured by the human body, which influence pleasure and memory, motor coordination and the perception of time. There are endorphins, which look like opiates and also ease physical pain. There is GABA, which specialises in encouraging neurons *not* to fire and so contributes to tranquillity and counters anxiety. If we add adrenaline, which gives momentum, and serotonin, which (among so many other things) contributes to self-esteem, it's easy to see that we have here a true chemical arsenal of mass satisfaction [p20].

Statistics show that because happiness is linked to wealth, it ruthlessly divides rich and poor countries. Countries also become happier when they become wealthy and unhappier when their economy takes a downturn. We are reminded by the gossip columns and glossy magazines that money and happiness are not always positively linked. Even so, it's clear that a healthy bank balance generally helps improve happiness.

And yet we also know that happiness is relative, especially since Dutch sociologist Ruut Veenhoven inaugurated the World Database of Happiness, a collection of tens of thousands of scientific studies on the subject. Those who are reasonably rich do not become happier by increasing their available wealth. Thanks to the hut and two pigs he owns, a Caribbean islander could be happier than a middle-class European who yearns for a house, a car and a lawn like the people next door. Psychologists talk of a hedonistic wheel, the wheel being the one in a hamster's cage, and hedonism the philosophy that sees pleasure as the purpose of every human action. This is the process through which, once the excitement of novelty has worn off, we wish for more. So you see, being content – or rather appreciating what we already have – is a rather effective trick for cheating a dissatisfied brain [p229]. But there's even more we can do.

They say that happiness is linked to pleasure, but also to

participation (passion for what one does), social relationships (family and colleagues), the sense of belonging (to a country, a volunteer organisation, a religion), and to one's own achievements (success obtained).

As ever, happiness depends on nature (chronic unhappiness is often genetic), and on culture, but also on life events and their continuous cycle of cause and effect. Common sense is very clear on this: 'If you reach your goals then you'll be happy.' And yet numerous studies show that it's only if you're happy that you'll reach these goals.

Evolution has added semi-automatic mechanisms to its operating system to facilitate happiness. An entertaining 1988 study provides an example. The subjects of the experiment were asked to say how funny they found various cartoons while holding a pencil in their mouths at the same time. The first group had to hold their pencils between their lips (forcing a sullen face), while the other group had to hold it horizontally between their teeth (forcing a smiling face). The result? You've guessed it. The mere fact of stretching the face muscles was enough to make the same, identical cartoon stories funnier.

If all we need to improve our mood is to use our face muscles, then how powerful is the 'trick' of positive thinking [p129]? One of the main characteristics of cognitive control, the brain's ability to adapt our behaviour to circumstances in real time, lies precisely in the ability to push away negative thoughts in favour of positive ones. This is an ability which can always be learned if absent, and therefore added to our anti-unhappiness defence system [p174].

Furthermore, since every brain has a built-in mechanism that keeps the hedonistic wheel turning, why not make it turn in the opposite direction and feel grateful for what we already have [p224]? Incredible as it sounds, there's always someone who's *far* worse off than we are and comparison with them, ungenerous as it might be, can activate dopaminergic results.

Oddly, generosity itself also stimulates a brain reward. And, according to some studies, meditation has the ability to activate (and so develop long-term) the left prefrontal cortex. It's also a well-known fact that physical exercise increases endocannabinoids and that sexual exercise – not indispensable to happiness and yet very useful – gives us dopamine, oxytocin and various endorphins [p20].

Finally, just to cover everything, there's also temperature to consider, since it has been proved that people are happier where it's cold than where it's very hot... No, that's not just a bad joke. It's precisely what could be inferred from reading the 2017 United Nations World Happiness Report. Out of the 155 countries tested, the three happiest countries in the world are Norway, Denmark and Iceland. The three unhappiest are Tanzania, Burundi and the Central African Republic. Naturally, this statistic is not based on temperature but rather on GNP per capita, social security, life expectancy, collective generosity, the perception of corruption and the freedom to choose one's lifestyle. The global level of all these indicators has increased over the past century. Although there have been highs and lows, the world happiness average has also increased.

Considering its positive effects on the cardiovascular system, on family relationships, on work and on life in general, we suggest you never neglect your conscious pursuit of happiness.

6.3 CONSCIOUSNESS

Consciousness is the easiest thing in the world. We've inherited it without any effort on our part, it is the companion of our every waking moment, and we find it perfectly natural that it should go on stand-by during sleep and then switch itself back on as soon as required.

Consciousness is the most difficult thing in the world because we don't know what it is. What's worse is that we can't even define it. It's the ability to perceive and experiment. It's subjectivity. It's the awareness of oneself and of one's environment. It's thought. It's free will. It's the control panel of the mind. It's all that and much more.

Consciousness is the brain's most mysterious characteristic. So mysterious that it attracted the attention of theologians and philosophers long before scientists got interested. The true nature of consciousness has been fiercely debated for centuries, especially since the seventeenth, when René Descartes identified his famous mind–body dualism. On one hand, I think, therefore I am: consciousness undeniably exists. On the other hand, it doesn't seem to have a physical existence, cannot be described or observed except by the brain that experiences it. Consequently, as was often the case with things that were hard to explain, it could only be a supernatural gift.

The proximity of consciousness to the concept of a soul has ended up cloaking it in taboo. Scientists have long been reticent about examining this topic in depth – also because it's impossible to study it in a lab. It has neither mass nor speed, so cannot be measured.

One who didn't shy away was Francis Crick, who jointly discovered the double helix of the DNA and who, during the last years of his life, led a scientific investigation into the mystery of consciousness. In his book *The Astonishing Hypothesis*, which he wrote at the age of 78, he suggests the 'astonishing hypothesis' – that 'a person's mental activities are entirely due to the behaviour of nerve cells, glial cells, and the atoms, ions, and molecules that make them up and influence them'. Nowadays, this is no longer surprising. Body and mind seem like two separate entities, but have actually been proved to be one and the same.

Even so, the mystery remains impenetrable. According to the Australian philosopher David Chalmers, the consciousness dilemma should be divided into a 'simple problem' (how the brain can produce memory, attention or thought) and a 'hard problem' (how one and a half kilograms of biological jelly can transform electrochemical information into qualities, like the 'yellowness' of yellow or the sourness of a lemon).

But are we sure that we need a complicated brain like ours to generate any level of consciousness? It is now widely accepted that primates that are evolutionarily closer to *Homo sapiens* (chimpanzees, bonobos, gorillas and orangutans) possess a certain degree of self-awareness, and the same is true for mammals such as dolphins and elephants. And there is more. In 2012, a prominent group of neuroscientists signed the **Cambridge Declaration**, which concludes: 'The absence of a neocortex does not appear to preclude an organism from experiencing affective states. Convergent evidence indicates that non-human animals have the neuroanatomical, neurochemical, and neurophysiological substrates of conscious states along with the capacity to exhibit intentional behaviours. The weight of evidence indicates that humans are not unique in possessing the neurological substrates that generate consciousness.' The term **neural substrate** refers to the parts of the central nervous system that take part in an action or emotion. Therefore, all mammals, and also birds, have a consciousness, albeit in different gradations.

Among the countless theories around, straddling science and philosophy, it's worth highlighting the **integrated information theory** put forward by Giulio Tononi, an Italian neuroscientist, and Gerald Edelman, American winner of the

Nobel Prize in Medicine. This theory is complex and deeply rooted in mathematics. In a nutshell, it says that 'a physical system is conscious in as far as it is capable of integrating information'. Think about it: your entire cerebral experience is a mosaic of information bytes that arrive from outside (visual, sound, tactile) and from inside (thoughts, feelings), and yet are totally inseparable. Consequently, the substrate of consciousness could be a system made up of various elements of information: the more a living species is capable of integrating them, the higher its level of consciousness.

We would like to confirm that the operating system installed in your encephalon [p10] gives you potentially the most integrated model of consciousness available to date, fully able to provide you with a Self.

6.3.1 Self-Awareness

Your brain knows it exists. It knows it is separate from other people in the four spatial–temporal dimensions. And it has the damned need to feel important, even if this means cheating a little.

This, in summary, is consciousness, self-awareness and self-esteem: three concepts that are deeply interconnected.

You're bound to be familiar already with self-esteem. It's a semi-automatic program that inserts into your thoughts sentences like 'I'm so great', 'I'm an ace at this' and 'at least this is something I'm good at'. It was introduced into the operating system in order to tackle unforeseen events (which were frequent in the case of our distant ancestors) and to support the motivation apparatus [p154].

Psychology requires that your brain should have a good reputation in its own eyes. It may sound odd that the brain should believe in such self-suggestion,* but that's precisely what happens, on both a conscious and a subliminal level [p140]. This is probably an effect of the central nervous system's obsessive need to predict the future [p60] and provide a reassuring projection of its own abilities while in the process of tackling the daily challenges, large and small, of living in one of the planet's metropolitan jungles. One study has shown that self-esteem lights up the primate brain right where the central portion of the prefrontal cortex (which deals with self-awareness) connects to the ventral striatum (which manages motivation and reward). According to another study, the less activated a certain part of the frontal cortex is, the more the brain is inclined to wear rose-coloured glasses, as though the rationality knob had been turned anti-clockwise. 'I'm exceptional, above average,' you think, and the serotonin levels rise [p24].

'Positive thinking', often celebrated by songs and cod philosophy, truly does have a positive impact on the brain because, by projecting an ideal image upon the imaginary stage of future events, it gives consciousness the strength to face them. A 'positive' brain foresees happiness, good health and success for itself, and feels the strength to overcome difficulties. On the other hand, a central nervous system that's tuned to negativity foresees its own inadequacy, and is worried and stressed. No wonder some of these 'forecasts' then manifest themselves in reality, during job interviews, for example.

We're not talking about black and white, but an entire chromatic spectrum. Each and every brain is unique and so are its positive and negative tendencies in handling the various fields of human activity. The range goes from having exaggerated

---

*   This phenomenon is especially peculiar if we think that, on the contrary, self-tickling doesn't work.

self-esteem (which coincides more or less with narcissism, a common error in calculation) [p186] to having such a low level of self-confidence that normal, everyday life becomes a living hell.

Many psychological studies have proved that human brains tend to see things as 'rosier' than they are: this is a peculiarity of the self-esteem program, evolutionarily installed to encourage happiness [p122]. In actual fact, in the psychological gradation that goes from narcissism to nihilism, it is the brains that sit in between the extremes, capable of a sort of balance between inner confidence and a belief in the reality of world's hardships, which manage best.

Self-esteem is a semi-automatic program, in the sense that it can be adjusted. All brains may experience depression at one time or other in life. But, except in chronic cases that border on the pathological [p202], inner strength combined with sociability [p132] yields excellent cures. There is always something positive to think about. The thought 'I've lost my job but I have two children who are fine and I own my house' is just a banal example of how the brain can be cheated. And the cheating works. If need be, find the formula that suits you best and try to respond to negative thoughts promptly, *every time* they come up.

Having said that, balance in levels of self-esteem depends greatly on the installation phase of the program [p79], during pretty much the whole childhood stage. It's not about telling a young brain how good, special and clever it is. Self-esteem is built when one feels accepted; capable but also operational when faced with daily educational, physical and social challenges. If you have already mixed your genetic heritage with that of another person, or are about to, you must be aware that a correct installation of the self-esteem program in a new human being depends, at least in part, on you.

Although self-esteem is an important part of self-perception, there is more. There is the physical sensation of having a body separate from others and, especially, the underlying awareness of one's own individuality, or **self-awareness**. The latter, after centuries of philosophical debates, is at the heart of the most relevant question of all today: is the artificial intelligence of a calculator and its algorithms going to evolve to the point of reaching self-awareness? For the time being this is still an open question, but it is a difficult question because it is at the crux of human experience [p242].

Your *sapiens* brain shares the ability to recognise itself in a mirror only with the brains of chimpanzees, gorillas, elephants and dolphins. However, no other animal in the world shares the ability to project itself mentally to a future time and place, to project words and actions and to take responsibility for them.

The Self is a cerebral construct that holds the strings of an avalanche of real-time information – and this information gives that sense of uniqueness that helps maintain balance and personality. When this uniqueness is interrupted or malfunctions, we get pathological cases of dramatic depersonalisation.

'The wise man' – we read in a passage of Buddhist literature – 'does not let himself be swayed by the eight winds: prosperity and decline, honour and dishonour, praise and blame, suffering and pleasure'. A solid perception of oneself is a good initial armour to tackle not only life's 'lows' but also its 'highs'. It's the passport of our own social identity, which we present at the border with empathy.

6.3.2 Empathy

Empathy arrived in the world 65 million years ago, when a huge asteroid collided with what is now the Yucatan Peninsula in Mexico. It is estimated that three quarters of animal species were erased by that historic event, which spelt the end of the Cretaceous Period, as well as the dominion of the ruling species: the dinosaurs.

This is how mammals came forth. Over the following 43 million years (the period known as Paleogene), starting with something vaguely resembling a mouse, they diversified to the point of dividing the hereditary line into different orders, such as Chiroptera (bats), Cetacea (whales), Perissodactyla (horses) and Primates (a species of which being *Homo sapiens*). Consequently, the dominion of eggs was replaced by that of placentas.

And a totally different parenting style.

The egg can look after itself and no sooner does it hatch than the newborn reptile is ready to operate in the world. In the case of mammals with placentas, however, the young offspring needs care, protection, warmth and food. It also needs to be taught quite a few things, varying in number depending on the species: no species is schooled as long as *Homo sapiens*. The limbic system has evolved in order to manage the emotional signals necessary for a new, emerging sociability [p42]. And sociability requires the ability to understand, as far as possible, other people's needs.

This ability is called **theory of mind**. It means to be aware of the fact that other people's mental states, such as their wishes and intentions, exist and are separate from ours. This is not a negligible detail, since you don't have access to another encephalon and yet you take it for granted that it is full of thoughts like your own.

Take one step up from theory of mind and you get to empathy. That is, the entirely human ability to put oneself into the shoes of another brain. Or of several brains at once. Although chimpanzees have a rudimentary theory of mind and a certain degree of empathic ability, they are unable to conceive of things like 'I imagine Mary knows that I would very much like her to make up with Albert, but I wouldn't want her to then go and tell Sarah everything'. This is exclusively human stuff.

Empathy is not strictly connected to an area of the brain, there is no consensus on its definition, and many ways of cataloguing it have been suggested. It definitely contains a wide range of emotions, from sensing the thoughts and feelings of another brain to the desire to help and support it if the need arises. In some cases, one can experience the same emotions as somebody else, as though there were a wireless connection between the respective limbic systems. In many cases, one can feel the pain of a person known only through TV news coverage, but never actually met. And, in the darkness of a cinema, one's brain can even worry about the fate of a character who doesn't actually exist.

Empathy, a necessity in the mammal brain and a masterpiece of the primate brain [p50], is at the root both of the continuation of the species and the start of civilisation. The ability to transmit and receive sophisticated emotional messages, the faculty of being able to project oneself into the mental and even physical state of another living organism, are the semi-automatic functions of every human brain. With, naturally, the usual individual differences.

Everybody possesses the neural ability to represent an interior model of external reality. It happens while we're watching our favourite sports event on television, where the internal model of a spectacular athletic feat generates pleasure (partly thanks to the contribution of **mirror neurons**, considered responsible for the emulation process). It happens when we're watching a film (Indiana Jones in that room full of snakes makes us shudder) and it happens when we're in a bar with friends (any story triggers a neural construction of scenes, events and situations). It's in the rear part of the temporal lobes [p56] in particular that your brain reconstructs reality in its entirety: in order to understand someone, you use yourself as a model. This already constitutes a big step forward towards empathy.

Your brain is able to project in imagination not only into another place but into another time and a different identity, which is essential for being able to put yourself 'in someone else's shoes'.

Many studies confirm that empathy, along with a certain inclination towards language, communication and human relationships, is especially pronounced in female brains [p178]. The male model, on the other hand, seems more drawn to systems, machines and categorisations. This has led to a peculiar theory: that autism may be an extreme form of the male brain.

Those on the autistic spectrum – called that because it's a highly variable malfunction in terms of phenomena and intensity – have greater or lesser degrees of difficulty in understanding the emotional content of a facial expression, or the allusions in a sentence, to the point where the empathy module can be totally non-functioning. The theory of psychopathologist Simon Baron-Cohen (a Fellow at Cambridge University and a cousin of the comedian Sacha) is that there is a kind of hyper-masculinisation at the root of this pathology, although 20% of cases affect female brains [p201].

Empathy can be resisted. Doctors and nurses, for strictly professional reasons, must have found a way to turn down the empathy knob. Functional magnetic resonance imaging (fMRI) scans show that, when seeing a needle piercing flesh, a surgeon's brain is synaptically less active than average. This phenomenon is not exclusive to health professionals. A young soldier, for example, who couldn't bear the thought of killing, discovered that after being at war for three months he was no longer affected by it.

The adventure of humankind is not only characterised by empathy and collaboration – quite the contrary. History books suggest an absence of empathy, a competition for resources, that results in one conflict after another. This competition is not universally disastrous, in fact it sometimes makes logical sense. According to game theory – the mathematical model of conflict and cooperation between intelligent and rational 'players' put forward in 1944 by mathematician John von Neumann – there are zero-sum games where one wins and the other loses, like tennis and boxing, but also **non-zero-sum games**, where both sides gain something.

Social, scientific, medical, artistic, technological, economic and even political progress over the past five, ten or even 20 centuries is based on a multitude of small and large non-zero-sum games between empathic brains. The constant – and accelerating – spread of ideas, discoveries, techniques and solutions has contributed to a cultural evolution that's like a prosthesis, an elongation of the biological one. This would never have happened if, in the brain of the last hominid species left on planet Earth, consciousness had not appeared.

Looking at history from a height can change the view of the world.

6.3.3 Weltanschauung

That said, we each run into our perception of the global whole: ourselves, others, the universe, accumulated culture, beliefs, convictions, ideas, inclinations, values, ethics and all they carry within them. It's our view of the world, what Immanuel Kant first called *Weltanschauung*. We prefer – and not only for reasons of its intimate closeness to philosophy – to keep the name of this component of consciousness in German.

Weltanschauung, the mosaic of a myriad of ideas reconstructed by the brain, contains all its points of view, its philosophy of life. Within it, there is epistemology (its interpretation of the nature of knowledge), metaphysics (its thoughts about the fundamental nature of reality), theology (whether or not there is a purpose to the existence of the universe). Then there's also cosmology and anthropology, not to mention axiology: what is right and what is wrong?

The view of the world is rather blurred during childhood. In adolescence, it can be the object of great turmoil. Then gradually it becomes sharp and defined. As adulthood progresses – through the complexity of long-term cognitive development mechanisms [p162] – it can even become polarised, in the sense that it becomes harder to change one's mind or perspective. It does, however, remain markedly more flexible in brains that have elaborated and stored information all their lives. This is yet another reason why lifelong learning is recommended for all brains [p216].

Weltanschauung is in constant motion. Because of the brain's plasticity, it is influenced by the environment from the very first days of life [p79] and, in theory, never stops being influenced. New information and new events add and alter ideas – all under the more or less conscious influence of the cognitive biases which afflict all brains, yours included [p189].

Since the brain is equipped with the semi-automatic empathy function, it will still be easy for you to understand that the world view of a woman who was born in an archipelago in the Pacific Ocean, a Pakistani peasant on the 'contentious' border with India and a female lawyer working for Goldman Sachs in New York will be somewhat different. Is this just nurture? Or does nature – genetics – also play a role?

Take the predisposition towards novelty, for instance, considered by psychologists to be one of the founding traits of a brain's personality [p149] and deeply involved in the overall view of the world. It is, in fact, commonly associated with the left end of the political spectrum, just as its opposite is associated with more traditionalist conservative thinking. Naturally, there is no shortage of studies about the brain and its political inclinations. And it seems in general that liberals have a slightly larger anterior cingulate cortex [p49] (the part of the frontal cortex above the corpus callosum), while conservatives allegedly have an atrophied right amygdala.

Even though there may often be a genetic component, one thing is certain: *Weltanschauung* is passed on first by the family, then by friends and school. Later, it is influenced by social interactions, studies, employment, television and the bedlam of information on the Internet. Reinforced by the so-called 'confirmation bias' (the psychological tendency to search for information or friendships likely to confirm pre-existing beliefs), it takes solid root in the brain [p189]. This is because, as we know, the brain is obsessed with predicting the future [p60], identifying recognisable models [p102], and also cataloguing, i.e. putting everything in specific pigeonholes.

Words like 'immigrant', 'black' or 'gay' promptly recall an associated category in every brain. In our example, the categories pre-set by an extreme right-wing brain and an extreme left-wing brain (not to mention the infinite gradations in between) will be markedly different.

World view includes morality. Even though your brain does not possess a 'morality centre', the inhibitory aspect of morality seems connected to the prefrontal cortex and the amygdala, the headquarters of rationality and impulsiveness. This is backed up by fMRI scans of brains catalogued as 'antisocial, violent and psychopathic', which show that their inhibitory mechanisms don't work well at all.

This is where, morally speaking, things get tangled up. If antisocial behaviour is caused by a specific brain malformation, should we really condemn it? If a particularly grave act is connected to genetics and/or a terrible childhood, should the brain that conceived it be locked up behind bars? If the testosterone levels of the defendant were so high that they promoted violence, do we solve the problem with a death sentence?

Now the synaptic reaction you experienced at the moment of reading the previous paragraph can give you a sample of your *Weltanschauung*. Whatever the moral judgement your brain readily expressed on such a delicate topic – the biological boundaries of guilt – we must stress the fact that neuroscience is constantly finding good reasons for reviewing the entire international judicial system, starting with countries that still have the nasty habit of wielding capital punishment.

Humans who populated the world before we knew that the Earth was round, or that life is reproduced by using exactly the same secret code, were bound to have a very different world view. Now there are world views that are shared and recognisable across various human cultures to the point of becoming the butt of jokes ('An Englishman, an Irishman and a Scotsman walk into a bar...'). Who knows? – perhaps during this century, which is so globalised and interconnected, within a generation or two a new shared *Weltanschauung* will sprout. One that is truly universal.[*]

---

[*] It would be genuinely useful to solve those two or three little problems on a planetary scale, starting with global warming, which risks capsizing the thermal balance of the Earth. In our experience on this front so far, 'local' versions of *Weltanschauung* appear not to produce the desired political and economic effects.

6.4 BEYOND CONSCIOUSNESS

Beyond consciousness is where the large majority of brain activities dwell. What you are able to perceive, even when you pay attention to all of your senses, represents just a tiny part of the underlying neural workings. This is easy to accept when we consider the structural complexity of the central nervous system. What your brain may find hard to digest is that even behaviour – the most conscious thing there is, since one's own actions and words are always in the foreground – operates mainly beneath the threshold of consciousness.

The concepts of unconscious and subconscious have become popular since the work of Sigmund Freud, who used both interchangeably before focusing on the former. The latter is often used loosely in non-scientific contexts. So, to avoid confusion, this manual will refer to **subliminal activities**, from the Latin 'beneath the threshold' of consciousness.

Freud reached the conclusion that subliminal brain activities influence behaviour. But the therapeutic solution he proposed and practised – digging into an unconscious past populated by strange feelings towards one's parents – is somewhat questionable. Partly this is because psychoanalysis is not an exact science (its success depends on a single patient and a single therapist), but even more it is because the conscious brain doesn't have access to the subliminal brain. Contrary to what is commonly believed, a brain area specialising in the management of sub-consciousness activities doesn't seem to exist, either. This is something Freud, who died in 1939, half a century before the invention of fMRI, had no way of knowing.

Nowadays, the classic concept of consciousness – the notion that all mental processes are accessible to the brain user

– is dead and buried. Vision, for example, a sense we consider infallible, can be cheated by a plethora of optical illusions [p102]. The brain function of memory is largely based on processes that are beneath consciousness, and implicit memory (writing with a pen, riding a bicycle) is based entirely in this realm [p64]. An extraordinary ability like language, which we take for granted, works thanks to the fact that the two systems, conscious and subliminal, work in tandem to produce a sequence of phonemes, or units of sound, that dwell beneath the consciousness level and manifest themselves as conscious thought, one word at a time.

The conscious system operates in **serial mode**, i.e. with a single series of operations in sequence. It's no coincidence that, as you know, it's hard to think of more than one thing at the same time. The system beneath the consciousness threshold, on the other hand, processes a far larger amount of information – from emotions to memories – in **parallel mode**, that is, by integrating them all together simultaneously. Naturally, the conscious and subliminal systems operate as one – and this is precisely the feeling you experience.

The important thing is to know that when you choose which house to buy, which share to sell or even which person to marry, rationality is not your only available resource for making the correct decision. Or the wrong one.

6.4.1 The Reward System

If, as we speak, you are experiencing the awareness of being, it is essentially thanks to two factors:

1. Millions of your ancestors survived to reproductive age.
2. They all reproduced.

If only one of them – somewhere in a genealogical tree that's far more ancient than the human race – had not defended himself against dangers, had not fed and taken care of himself, or if, one fine day during his life, he had not had a sexual urge, you would not be here.

If, as we imagine, you're glad about that, then you must know that you have just felt a discharge of dopamine. It's the same molecule that distances people from risk while drawing them closer to food and sex.

To draw closer or distance oneself. Accept or avoid. All the brains in the world have evolved according to this automatic and subliminal model for mundane reasons of survival: drawing closer to food or moving away from it, accepting it or avoiding it. This model is called the *reward system* because it is activated by producing brain stimuli that 'reward' a specific action or, on the contrary, discourage it. This system works in tandem with the learning module [p162] (which can reinforce the motivation to repeat the action), as well as the memory module [p64] (in order to remember a lesson for the future). All this is vital for modern society: without the reward system, the hamburger and chips, the orgasm, the Internet, bets, cocaine and shopping would not be the same.

The currency generally used by the reward system is dopamine [p25]. The axons of dopaminergic neurons leave from the midbrain (the ventral tegmental area, VTA, to be precise) then spread until they reach either the *nuclei accumbentes* [p48] (the so-called **mesolimbic pathway**) or the prefrontal cortex (**mesocortical pathway**). They are called dopaminergic because they can release dopamine to the synapses. The VTA also sends dopamine to the amygdalae (emotions) [p44], the hippocampus (memory of events) [p45], and to the striatum (learning), using other pathways [p48]. There are still more pathways that go from the *substantia nigra* [p40], also in the midbrain.

Without your being even minimally aware of this tangle of electrochemical events, when dopaminergic neurons fire, you experience a pleasant sensation that prompts you to 'draw closer', and of which you will keep an associated emotional memory and a more or less conscious awareness. Conversely, if, for example, you smell food that is off, your VTA registers the fact that it's something you need to distance yourself from and promptly tells the *nuclei accumbentes* to generate disgust, making sure it also informs the hippocampi not to approach blue-coloured food ever again.

The fact that nothing would be the same without dopamine was revealed to us by wretched lab mice (which share about 97.5% of our *Homo sapiens* genetic heritage). When they are given a substance that blocks dopamine receptors, they stop eating, potentially until they die. No sooner are the receptors reactivated than they start feeding again as though nothing had happened.

Chimpanzees, on the other hand (whose genome is 98.8% the same as that of humans), have shown us that calling dopamine 'the pleasure molecule' is not, strictly speaking, correct. Many laboratory studies have proved that, contrary to what we thought, neural reward does not 'fire' at the end of the action but *before* it is performed [p60]. Or, rather more precisely, in the first instance, dopamine floods the brain of a test animal when it discovers that, by pulling a lever three times, an edible reward rains down from the sky. Thereafter, once it has learned what to do, the neurotransmitter floods its brain to trigger the *ex-ante* action and not to reward it *ex-post*. This discovery has given rise to the new idea that dopamine is concerned with desire more than pleasure.

Experiments on chimpanzees, macaques and other primates have added another new piece to the reward system puzzle. If, after learning the trick, the animal gets used to self-administering rewards on demand, when the game stops (thanks to a sadistic scientist) it becomes agitated. One might say that it manifests symptoms of longing, which means that the activation of the dopaminergic path has become a habit.

Habits are a kind of circle that repeats itself, and are very useful in daily life. You've put all your attention into learning the sequence of operations to turn on the engine and start the car; now you perform the same procedures without even noticing. But habits also make the economic system go round. If just walking past that ice-cream parlour is the signal that triggers the impulse to grab a cone, it relies on habits incorporated into the subliminal reward model.

Thanks to this model, advertising, marketing and the psychological study of consumers generate profits on a planetary scale: if the customer's VTA is activated *before* taking that brand of toothpaste from the shelf, it means the TV commercial has hit its target.

Sometimes habits take the upper hand and become priorities. The flip side of the often beneficial coin of neuroplasticity is that one can become dependent on fatty foods, Internet porn, video games, roulette, alcohol or the shopping trolley, sometimes with unpleasant repercussions. 'Wanting' turns into a need so pressing that it damages your health, reputation, job and family. At this stage, habits are called **addictions** [p193]. Scientific research on why all brains don't have a tendency towards developing these addictions is ongoing, but recent studies have identified the psychological and neuroplastic pathways we should follow, even if only to eliminate small, undesirable habits [p193].

In the final analysis, the entire reward system is a complex kaleidoscope of experiences that are both present in the immediate moment and longer-lasting, because they involve the memory of the past. They are also projected into the future – after all, they help you make decisions you no doubt consider rational, conscious choices. We regret to inform you that these, too, are mainly subliminal.

6.4.2 Free Will

A little while ago, instead of reading your e-mails, you decided to pick up a book. You read a few pages then decided to stop right here (even though you could have chosen somewhere else). You could equally have suddenly closed the book and switched on the television but, instead, you chose to continue as far as these words.

Everyday life appears to be a sequence of free choices. Unfortunately, according to many philosophers and scientists, all this great freedom is simply an illusion. Free will does not exist. By reading this book and these exact words, you have simply followed a predetermined script.

This kind of statement is at odds with daily experience but, assuming that free will really is an illusion, it is also at odds with the doctrine of cause and effect we are used to – so much so that it throws doubt on the concept of responsibility which is at the root of all the civil and penal codes in the world, as well as many religious doctrines, since rules and command-ments are based implicitly on readily available free will. No wonder it's a topic that ends up exercising many brains.

In case this includes you, we would rather not leave you on tenterhooks. In order to placate the stress hormones currently circulating inside you, let's say three things. First, that free will is a mystery without any definite solutions; second, that much depends on its definition; and third, that it's reasonable to believe that you have at least a certain amount of free will. But now let's take things a step further.

There is no philosopher in history, from Aristotle to Kant, who has not said his penny's worth on free will. The ideas that have been in circulation throughout centuries of reviews and reflections are immensely complex. To put it extremely simply, we could say that free will is not compatible with **determinism**, the notion that for every event there is a cause that can only provoke this very outcome. This was originally an Ancient Greek concept, but it has found fertile ground since Isaac Newton proved that the universe moves according to a series of relentless universal laws of physics. The human body (brain included) is made up of atoms: the same ones that form the deterministic macrocosm in which you currently live.

However, almost a century ago, it was discovered that in the microcosm of these atoms, the rules of the game change surprisingly. In the picoscopic world of **quantum mechanics** (a hydrogen atom is 25 picometres large, 25 millionths of a millionth of a metre), 'objects' like photons and electrons have a double nature of waves and particles and are, above all, governed by uncertainty. At the dimension of sub-atomic particles, deterministic certainty is replaced with probability.

That is, until 1985, when psychologist Benjamin Libet invented an experiment to measure the lapse of time between volition and action. After connecting his subjects to an electroencephalogram to record brain activity and an electromyograph to measure muscle activity, Libet asked them to bend their wrists at specific intervals, and to signal the exact moment when they decided to do so. He discovered something surprising: in the frontal cortex area responsible for preparing movements, there is a cerebral activity, called **readiness potential**, that occurs 350 milliseconds before the subject claims to be conscious of his or her own action. For many people, this confirms that voluntary acts and decisions start subliminally, beyond the threshold of consciousness. There is no room for free will.

At this point, ethical doubts about the biological boundaries of guilt [p136] could be dismissed, but the belief in a *point blank* absence of free will cannot be reconciled with a civilisation based on civil and penal responsibility. The notion that there is no free will is demonstrably damaging. In a 2008 psychological study, two groups were read two different quotations by a famous scientist: the first a general statement on consciousness and the second one that categorically excluded the existence of free will. Then a game with a money reward was organised, in which it was easy to cheat. Those who felt liberated from the burden of free will were 45% more inclined to behave unethically.

As luck would have it, among determinists, libertarians, incompatibilists and revisionists (these are just some of the trends of thought that come up against one another in this debate) there are also **compatibilists**. That is, those who believe that there's room for free will in the deterministic universe.

British physicist Roger Penrose suggested that quantum mechanics, which regulates matter at a sub-atomic level, could explain the existence of consciousness as well as free will. There are those who dispute this theory (by claiming, for example, that it is not compatible with the operating temperature of the brain) and those who embrace it to the point of turning it into fantasy theories about the 'quantum mind' [p206].

Philosopher Daniel Dennett, author of the book *Freedom Evolves*, claims that 'We are freer than our parts [...]. Without adding anything mysterious', in as far as the difference isn't in physics but in biology. A billion years ago, there was no trace of free will on earth. Then, gradually, evolution added to the abilities and skills of living organisms, which, in *Homo sapiens*, reached a peak when they acquired something that was unique in the animal world: being able to foresee the future consequences of one's actions. It's the basis of morality and social living.

Think about it. If someone asked you, 'Why did you do that?' you would be able to reply in a precise, rational manner. The brain commands actions represented to it at a conscious level, but which are influenced on a subliminal level. It's possible that the threshold between these two levels may vary. If you express yourself in a language you're still learning, you'll need a considerable effort on the conscious level to connect verbs and conjugations. In your mother-tongue, however, it's your unconscious that takes the upper hand. Imagine a saxophonist in the process (which comes naturally for him) of improvising on the harmony of a song: as the notes he plays flow, he could be taking as many as five or ten 'decisions' per second. He would never have enough time to do this on a conscious level, but only by drawing on the musicality he has integrated subliminally in his encephalon. There is no spirit guiding him, no destiny controlling him, and everything is done by a brain – composed of a part that he is aware of and a part of which he is unaware – that works perfectly well as a single entity.

Of course, as the eminent biologist Jerry Coyne says, it could also be perfectly true that 'free will is an illusion so convincing that people simply refuse to believe that we don't have it'. As we have seen, however, not believing in it is neither good for health or for society.

'Do you believe in free will?' journalist and writer Christopher Hitchens was once asked in public. He promptly replied, 'I have no choice.'

6.4.3 Personality

The luckiest man in the world is called George. He decided to put an end to it all in 1983. He was nineteen years old, a good student, and felt certain that there was no escape from his condition: his obsessive–compulsive disorder [p204] forced him, among other things, to wash his hands dozens of times a day. His social life was compromised, and his inner life was all over the place. One day, he took a gun and shot himself in the head. The doctors operated on him. He survived. But now he was finally free of his obsessions: the bullet destroyed the neural centre of his disorder, actually turning him into a different person.

George was luckier than Phineas Gage, who, having survived a much larger perforation of a frontal lobe, became the victim of a drastic deterioration in personality [p53], although it was thanks to Gage that scientists in the mid-1800s began to suspect that there was a direct link between biology and behaviour.

Behaviour, which originates from the sum of consciousness plus all the underlying subliminal chaos, is the expression in time and space of the individual personality. And personality is that unique imprint of every human mind, made up of an indefinite blend of values, memories, social relationships, habits, passions and interests.

Psychologists, who study behaviours, have produced numerous classifications to try and pigeonhole the range of personalities exhibited by the human race. Perhaps the best-known and most widely used approach divides people into five categories, assuming that all brains fit somewhere on the scale for each category. **Openness**, for instance, defines a predisposition to new things. You could tend to run away from any new experience, be it intellectual or sensory. Or else you could think that anything you've not yet done, seen, heard or tasted should be done, seen, heard or tasted as a matter of principle. It's most likely, though, that your brain is located somewhere in between these two extremes.

Conscientiousness divides those who program and organise their lives in minute detail, respecting deadlines as though these were laws, from those who leave everything to chance, never worry about anything and don't give a damn about deadlines. Needless to say, no brain in the world is so exaggerated: most belong to the intermediary shades of the conscientiousness rainbow.

Extroversion is the most obvious example of the possible gradations of personality, perhaps because in many people's view it's most evident and 'measurable'. The most extrovert brain is not the first to show up at a party or a dinner, but the one who will attract attention to itself until the final moment. On the other hand, the introvert brain by definition is not the one who stands in a corner, holding a glass, and is the first one to leave – it's the one who doesn't actually show up at the party at all.

Agreeableness, which depends on relationships with others, can be more subjective and variable. Gradations range from the brain that will do anything to be loved by everyone to the one that will do everything to be hated.

Finally, there is **neuroticism**, the most complicated category from every point of view, where 'high-scoring' brains tend to experience negative emotions like anxiety, fear, anger, frustration, jealousy, guilt and so on.

These five categories, which some affectionately call the 'Big Five', risk excluding some important shades of the much more complex mosaic of human expression, even though they do help us to understand its extreme variability.

There are other pressing questions: what determines personality? Where does it reside? What is more important, nature or nurture?

There is a tendency for the same brain to behave differently at 60 than it did at 20, often narrowing its openness to new things. But even within the space of a single day – because of events, environment or hormones in circulation – the degree of extroversion or conscientiousness can change dramatically. And yet, as you will also have noticed, other brains appear essentially stable in what we call temperament or character: the child with a well-developed sense of humour (now here's a shade that escapes the Big Five) tends to be equally entertaining as an adult. The reason for this lies in the fact that the personality is marked by nature as well as genetic heritage and the culture of the family environment. But not in a consistent, balanced way.

Although DNA and the environment have an equal impact on the creation of intelligence, for example, things change when we talk about personality.

Many neurological and psychological studies show that the degree of extroversion and neuroticism depends more on the mother's chromosomes than on what she does or says to the child. In other words, 'nature' shapes personality much more than 'nurture'.

This notion, put forward and argued in 1997 by psychologist Judith Rich Harris, contradicts the previous view, which was based on common sense: that exposing children to violent behaviour produces a violent child; extroverted behaviour produces an extroverted child.

And yet everything tends to indicate that this new intuition is correct: however much a parent strives to shape their children's personality, the efforts are next to useless. It's one of those rare cases where the chicken comes before the egg. If a parent is sporty, the children will be athletic. If a parent enjoys reading, it's very probable that their child will enjoy reading. More than a matter of imprinting, however, it is the fact that the respective tendencies are stamped in the genes of the mother and the father, and inherited – 50% from her and 50% from him – in the child's genome.

Many studies, mainly on twins (who share genetic inheritance as well as a bedroom), confirm this theory. But there is more. A recent complex study conducted in Denmark examined over 14,000 adults adopted during childhood and, after comparing their criminal records with those of their adoptive and biological parents, showed that tendencies to criminal behaviour have a hereditary component. Another study conducted in 2015, once again at the University of Copenhagen, as well as Georgia and Texas, investigated (through MRI scans) the brain anatomy of 107 chimpanzees, all classified according to their respective temperaments. It transpired that those with a dominant personality had more grey matter in the right prefrontal cortex. Those classified as open and extroverted, on the other hand, had more in the anterior cingulate cortex of both hemispheres.

Those who avoid social gatherings are hardly likely to be the heart and soul of the party, and those who like to take on responsibilities, large and small, are not likely to start practising absenteeism. However, the brain is plastic. The personality is influenced by the reward system, which can often be corrected. A person who isn't very conscientious can choose to improve their reliability when faced with one of life's obstacles, or when they decide simply that it would be more convenient. A person who is inclined to feed their fears (if he or she has not crossed the pathalogical line) can learn to moderate their neuroticism [p174]. This is provided, of course, that the people in such cases have a bit of knowledge of the brain's control panel and its intricate mechanisms.

7.0 THE CONTROL PANEL

THROUGH THE CONTROL PANEL, YOU CAN regulate the voluntary functions of your brain, like motivation, attention, learning, imagination and emotions. Except that there is a problem. Every brain is shaped by genes (the chromosomal heritage), epigenetics (changes caused by modifications in the expression of genes rather than the alteration of DNA sequences themselves) and memes (the nurture you have been exposed to since birth). Consequently, there are no two brains in the world that are identical, which makes it impossible to provide a manual that describes a control panel that is the same for everybody.

Luckily, in neural function there are common principles which allow any user to exercise a considerable margin of control over his or her own brain – not without a high degree of commitment, though: the voluntary functions of the brain require, by nature, a heap of voluntarism.

Without motivation, it's hard to maintain attention. Without attention, learning is impaired. Without learning, knowledge is not fuelled. Without dynamic knowledge, imagination doesn't take flight. Without imagination, you can't attempt – and there is no shortage of obstacles – to make rational decisions or solve more difficult problems.

But what do these brain properties have in common? Well, the fact that they can all be learned, improved and perfected. You can learn to learn. You can adjust the emotion control until you alter undesired habits. In other words, you can do much more than what many brains tend to appreciate.

7.1 MOTIVATION

Motivation is standard equipment in every brain. Yet the motives that drive you to the bathroom or kitchen every morning are quite different from the ones that encourage you to learn to play the guitar or speak Spanish. The first, in fact, are automatic and depend on stimuli, while the second contain explicit will, generated by the cerebral cortex [p51]. These are the ones that concern the control panel of your brain.

Procrastination, the most frequent enemy of motivation, did not originate in the digital era. 'Procrastination is always hateful,' Cicero wrote more than 20 centuries ago. According to some studies on this topic, 20% of the world's brains are affected by chronic procrastination, but we take it as read that the 'I'll do it tomorrow' argument is even more pervasive than that statistic suggests.

Thought works on two parallel cerebral systems. One is rapid, largely automatic and mainly beneath the threshold of consciousness. The other is slow, reflective and manifests itself through the 'voice' you hear in your head. The first is based on the primitive structures of the 'reptilian' [p37] and 'mammal' brain [p42] (like the cerebellum and the amygdalae). The second relies on the sophisticated structure of the 'primate' brain [p50], the neocortex. These regions are often at odds with each other.

It's the neocortex that decides that the time has come to join a gym, for both visible (your tummy) and rational (your health) reasons. So it moves your legs as far as the membership desk, where you pay for an annual subscription, provoking a strange sense of wellbeing triggered by dopamine [p25]. Strange, because the reward derives from the joy of finally having taken a healthy decision, even though it's almost certainly an illusion [p199]. In the United States (the data for Europe is not available), 67% of all those who join a gym never go there to use it. This is a typical case of the 'reptilian' brain taking over. The innermost layer of your brain is not only lazy: it hates change. It's wilful and bossy. The sight of a cake, of an attractive human being walking past, or the *ding* message alert on your phone immediately shifts its attention from the important things like work, studying or driving the car.

This does not mean that the 'primate' brain cannot command it to change, do something or pay attention. It does try. It even succeeds. But the other region, like a child, suggests a minute later that it's time to take a look at Facebook, or that it's not worth attending the conference today because 'it's glorious weather' or 'it's filthy weather' or 'there's a bus strike' or 'the bicycle has a puncture'. In the absence of strong motivation responding tit for tat, the primordial brain always wins. We figure you know exactly what we're talking about here.

Still, after agreeing that you can't get the primordial brain away from the helm, it's possible to cheat it by using the same mechanisms that make it work.

Emotional circuits. When all is said and done, motivation is switched on through a process of association: when we think of a goal, we bring out a corresponding state of mind. Remembering happy experiences from the past (memory is closely linked to the emotional apparatus) contributes to creating a positive mental state. Many great tennis players, so solitary in their enterprise, talk to themselves during a match (sometimes even out loud) in order to 'spur themselves on': evidently it works. Moreover, varying your work routine or adding a form of novelty or fun to repetitive tasks can be emotionally helpful.

Cognitive biases. It is psychologically easier to complete a task that's already begun than one yet to be started. It's one of the many cognitive biases [p189] diagnosed by science and it is confirmed by popular wisdom: 'a job well started is halfway done'. Moreover, dividing work or study into blocks gives the impression that it's all straightforward and manageable.

The reward system. Procrastinating means preferring an immediate reward [p141]. Motivation, on the other hand, concerns the future: it supports an action that *will lead* to a future reward, such as a diploma or a promotion. Being able to project a final result in the mind, even at a distance of years, is a characteristic exclusive to the human brain mechanism. Even just imagining reaching a distant goal triggers a reward. The same occurs with shorter-term objectives, which is why dividing studies or work into blocks allows one to experience the dopaminergic boost repeatedly in a single day. This way, the neocortex can consciously try to placate the automatic restlessness of the 'reptilian' brain.

Plasticity. The brain's tendency to be mainly motivated or unmotivated is due to its structure and the entirety of its connections. By studying the two kinds of inclinations through functional resonance [p234], Oxford University researchers have found traces of this difference in the communication between the cingulate cortex and the pre-motor cortex of the frontal lobes [p53], which is involved in the decision to take an action. The interesting thing is that maximum activity is not found in motivated brains but in unmotivated ones, as though their connections are so ineffective that they require even more energy to go from thought to action. In light of this, it is easier to understand why some struggle with motivation more than others. Still, in case your overall degree of motivation should demotivate you, be aware that plasticity [p69] is also operational in this case.

It's true that motivation and attention are necessary for triggering the required biochemical effects that herald the creation and consolidation of synapses. But, just as optimism or, at the other extreme, anxiety, alters the structure of the brain, the practice of motivation also has long-term, plastic effects.

The match between the ancestral, automatic and emotional part of your brain and the evolutionarily modern, slow and reflective one, is played out every day and at every hour. The outcome is in your hands, or rather, your frontal lobes.

ON	OFF
Planning and acting	Procrastinating
Knowing that there is a conflict between the 'reptilian' brain and the neocortex	Thinking you are in charge.
Knowing that motivation can be incentivised.	Passively accepting lack of motivation.
Using emotions, reward and plasticity in order to take charge.	Letting yourself be dragged down by lack of motivation.

7.2 ATTENTION

At this very second, you are looking at a book. Actually, to be precise, you're focusing your eyes on a sequence of signs that represent words that encode a meaning that your brain has learned to interpret. In simpler terms, you are paying attention to what you're reading.

And yet it's not as easy as that. At the same time you also

feel the weight of your body on the chair, the wind blowing through the window and caressing your skin, the smell of a roast wafting from the kitchen, traffic noise in the distance and the girl on the fifth floor practising her cello. And not just that. Besides the deluge of information raining on you from outside, inside you can also feel the flow of your thoughts overlapping with the words in the book. How is it possible that in the midst of this chaos, you manage to understand? How can you concentrate on the words and keep everything else out?

The attention system is an integral part of the human brain, starting from birth. It helps focus the mind on stimuli that require a response: from the cuddle that heralds feeding time to the cry of a wild animal that recommends running away. At present, in our post-industrial society, since intellectual jobs have outnumbered manual ones, attention has become an even more essential economic resource, a measure of your actual productivity. It's a shame that it's a limited resource, and the one that is put to the test the most.

The idea of multitasking derives from the analogy between the human brain and the computer. Just as a calculator is able to perform several tasks simultaneously, we think that the human brain born or brought up during the digital era can multitask between work, text messaging and e-mail, Facebook and WhatsApp notifications. But this is plainly nonsense. The brain is definitely able to arrange an appointment over the phone, listen to the radio and drive at the same time, but not without the risk of an accident. Not only is the capacity for attention limited but, while turning one's attention from one thing to another, one usually experiences a roughly half-second 'gap' – called an attentional blink – during which the system simply isn't working. And if one is travelling by car at 60 miles an hour or more, half a second can be vital.

Naturally, a distinction has to be made between automatic attention (turning when one hears one's name called) and voluntary attention (deciding to read a chapter of that book). The two things conflict: it has been proved that interruptions caused by phone calls, e-mails and other messages interfere negatively with a student's learning. It has also been shown that they contribute to lowering work productivity by up to 40%.

As though external distractions weren't enough, control of neural attention is sometimes also compromised by internal factors that are beautifully connected to our personal cerebral configuration. The existence of attention deficit hyperactivity disorder (better known as ADHD) was acknowledged only in the 1980s and diagnosed on a large scale in the 1990s. Three times more common in boys and with a significant genetic component, attention deficit is characterised by a lack of concentration, impulsiveness and, in some cases, a degree of hyperactivity that impairs learning. What used to appear in schools as an abominable behavioural trait is now classified – albeit with a few distinctions – as pathological. It usually disappears at the end of adolescence, but in 30% of cases it carries on into adulthood. According to some estimates, 2% of the adult population is affected.

The attention system is regulated by dopamine [p25]. Even though there is no individual part of the brain dedicated to attention, more intense activity has been registered in the frontal and temporal cortex [p56], where neurons start to scream. It's a figure of speech, of course: but just as we raise our voices to be heard in a noisy environment, some neurons seem inclined to raise the intensity of their messages precisely to filter through distractions. According to some studies, the brain also concentrates by synchronising the 'firing' rhythm of some neurons to the point that music education has been suggested as a possible way to alleviate the symptoms of ADHD.

The neuroscience of attention is trying to unpick this problem for more than one worthy reason: attention deficit causes accidents and social disparity, and is currently treated (even in children) with drugs that have significant side effects [p230].

As usual, your brain is able to learn – to concentrate better, improve its ability to pay attention and thereby encourage successive efforts – through the reward system [p141].

The first step in improving one's degree of attention is to disconnect for a couple of hours from the information flow that comes from the smartphone and tablet, our seductive weapons of mass distraction. It is clear that multitasking increases stress and does not produce the desired effect of doing more and faster: one actually does less and it takes longer.

As everybody knows, there's attention and attention. One can absent-mindedly pay attention to a football match while reading a book and, as soon as the commentator's voice rises, concentrate on the potential goal action. If, however, one wants to learn, one must always concentrate. Concentration requires an act of will, and it can be learned and improved. According to some studies, attention is linked to the ability to exclude irrelevant thoughts or impulses.

Obviously, nobody can maintain attention for an unlimited time. Many people have adopted a method called the *pomodoro technique*, because its inventor would apply it using a tomato-shaped kitchen timer (*pomodoro* means tomato in Italian). It consists of performing an intellectual task in 25-minute blocks, with a 5-minute break in between. After four blocks, which correspond to two hours, there is a longer, 20-minute break. The point is that to keep the attention going for four hours in a row would not only be difficult but counterproductive. The breaks help you to focus better.

'Attention economics', a concept articulated as early as the 1970s, has become prominent since the new industrial giants (Google, Apple, Facebook, Netflix, Amazon) began competing with traditional media to share the same resource: public attention, which, like all economic resources, is characterised by scarcity. One cannot pay attention to TV series, websites, newspapers, apps and video games for more than a certain number of hours per day.

As Cal Newport, Professor of Computer Science at Georgetown University, suggests in his book *Deep Work*, attention is the new key factor in competition. In his opinion, it's the brains that are able to mentally dismiss distractions that will be successful in the new economy. Not only because intellectual work will become even more prevalent, but because it seems that the capacity for attention – in the current information chaos – is decreasing on a world scale.

As usual, though, through conscious effort and repetition, the degree of attention can be improved. Some say that meditation can be of great help [p224]. As long as one does not glance at one's smartphone while meditating.

ON	OFF
Trying to disconnect from distractions.	Constantly multitasking.
Training the ability to concentrate.	Believing that concentration cannot be improved.
Keeping up the pace of concentration but with breaks.	Concentrating every so often.
Meditation.	Sensory downpours (parties, for example).

7.3 LEARNING

It is believed that your brain started learning when it was still surrounded by a liquid, muffled and semi-dark world. It is there that it began mobilising its first synaptic mechanisms, anticipating the sensory deluge of birth [p79]. From that point, it progressively turned into the most powerful machine in the world, a machine that specialises in learning.

Plants, too, learn in their way. All animals, albeit to different degrees, learn. But none like *Homo sapiens*, who has built upon this evolutionary characteristic the civilisation of which you are a part.

For the time being, at least, the human learning machine is also more sophisticated, flexible and powerful than machine learning, the automatised learning at the root of the growing technological wave called artificial intelligence [p242]. A one-year-old girl learns to balance in a way no robot would dream of. A three-year-old boy recognises a lorry no matter what it looks like, its colour or its orientation in space. A fifteen-year-old adolescent can write, kick a ball in mid-air and amply describe the problems of the world around them. And so forth for the rest of his or her life, where the sum of acquired, reinforced or modified information is encoded in neuronal memory through a *never-ending* process that contributes to designing personality.

Think about it: you are what your brain knows and is able to do.

Learning is the most crucial function of the human brain, from gestation to cremation. Individual users, families and nations know this perfectly well, albeit often only in theory. There is no better long-term investment than devoting time and resources to synaptic growth [p18], exercising action potentials [p12] and producing myelin [p32] for oneself, one's children or one's fellow citizens.

Now let's forget about cases like that of the British child Tristan Pang who, at the age of two, could read and do high-school mathematics and who, in 2013, at the age of thirteen, started university; or like Joey Alexander, the Indonesian jazz pianist who, in 2016, at the age of eleven, obtained a Grammy nomination with his first CD – the brain mechanisms of these natural prodigies are unknown. They have in common an extraordinary capacity for attention and a childlike motivation that turns books or a piano into their favourite game. But let's be clear about one thing: in the home where Tristan was born there would have been no shortage of mathematical books, and in Joey's home there would have been quite a lot of jazz records. If Mozart had been born in a village in Siberia, instead of in Salzburg with a harpsichord in the living room, musical history would have been different.

Let's leave that 0.0001% of special brains aside. The remaining 99.9999% have to work hard in order to learn. If you struggle to learn mathematics or play the piano, rest assured that this is perfectly normal. The crucial question is: do you like working hard or – on the other hand – do you hate it? Because it would be beneficial if you considered hard work a joy: it would mean that the long-term development of your synapses is working. As confirmed by the work of psychologist Carol Dweck, the brain becomes more intelligent if it thinks it can become more intelligent [p72]. Equally, the brain learns better if it is convinced that it can learn anything, without the unhealthy notion that talent is fixed by destiny.

Not only that. Studies of learning mechanisms show that in order to truly master a subject, language or musical instrument, you must, as much as possible (and without overdoing it) take your brain outside its comfort zone, the zone where it doesn't have to work too hard. In his book, *Outliers: The Story of Success*, Malcolm Gladwell talks about 'the 10,000 Hour Rule': whatever the human activity, you become an expert in it if you study or train correctly for 10,000 hours – on average 20 hours a week, for ten years. It's all a matter of time, true, but it's that 'correctly' that makes a difference. Let's take the child pianist. If, as a tiny child, after learning three songs and all the major scales, he had been content, and tried to improve only on what he knew without ever leaving his comfort zone, the 10,000-hour rule would not have yielded the same results. He would not now be living in New York with an EB-1 visa (the one the US grants to geniuses) in his passport.

Repetition over time is part of the learning game because that's how the memory mechanism works [p64]: neurons that fire together, wire together, says Hebb's Rule. It's the repeated use of the synapses that strengthens them. The crazy night spent studying before an exam can help us to pass the exam, but not to remember what has been studied long-term. Even the genius – albeit working less hard – must repeat the learning process again and again over time if he or she wants to become a true genius. This is a way of saying that, no, there's no point in your looking for shortcuts either.

It's not just the synapses. The entire system is so firmly rooted in the repeated and progressive use of information that not only neurons but also glial cells take part in it [p28]. The astrocytes keep axon activity under control and, if it becomes raised, command the oligodendrocytes to add more myelin to protect them and enhance the speed of the transmitted signal. There is a direct connection between the amount of white matter and intelligence, knowledge and experience. The learning machine *is* extraordinary. But it must be used. And used correctly.

We have talked about learning in generic terms but in

actual fact it's a huge topic. Learning a foreign language, such as Polish or Spanish, 'lights up' many areas in the brain. Motor coordination, when, for example, learning to swim or skate, involves other regions of the brain. Playing the flute or the accordion involves a mix of the two, plus others. It's as though there were potentially a linguistic brain, a sports brain, a musical brain and so forth. There is room for all three in your skull, and for many more.

Since, during childhood and adolescence, neural development goes through so-called critical phases [p79] when certain kinds of learning are facilitated (languages, for instance, at pre-school age), it's sensible to use the first period of your life to attend school and, at the same time, go to the swimming pool, take dance lessons and learn as much else as you can. However, this is where it comes down to families and, especially, governments, which manage public education.

There are as many education systems in the world as there are countries. Schools in Finland, which have been ranked first in the special classification devised by the World Economic Forum, are essentially different from those in Canada, Austria or Senegal. In general, however, the majority of school systems do not provide pupils with even minimal basic information on that very brain they need in order to study (in an Italian secondary school textbook, we counted nine pages devoted to the brain and twelve to the digestive system), and don't even take discoveries in neuroscience into consideration.

To begin with, your standard strict teacher and rigid exam deadlines trigger the production of cortisol [p27], the stress hormone. In the presence of fear signals, the most primitive functions of the brain end up opposing the more modern structures of the cortex, impairing how they work [p37]. In Finland, the first exam is not until the age of sixteen, the critical period appropriate for experiencing a little stress.

Naturally, even far away from Helsinki, there are wonderful teachers who do their jobs well, from kindergarten to university.

They are the ones who manage to make their teaching interesting if not actually fun, even if they are not aware of this being the only way to trigger the circulation of **dopamine**, which favours and strengthens synaptic connections.

They are the ones who step down from behind their raised desks and keep close contact with students, which is said to encourage the production of **acetylcholine** [p25], the attention neuromodulator, in their students' brains,.

They are the ones who bring new elements to the class (for example rearranging the classroom or using an original teaching solution), encouraging the spread of **noradrenaline** [p24], which encourages attention and also, in the long run, attachment to studying.

And then, of course, if the pupils cross the line, they can always resort to a dose of **adrenaline** by raising their voices or threatening consequences. Still, they are the ones who don't brandish threats as a matter of habit, otherwise **cortisol** would ruin everything.

Now the problem is that you have to be lucky enough to end up in their classes (or be born in Finland*). To solve the challenge of successful learning at an international level, we would need all the various education systems to be adapted according to the main discoveries in neuroscience.

Should you no longer be of school age, you may find these observations tedious. We recommend you think differently. Learning is the most crucial function of the human brain

---

* In Finland, children start school at the age of seven, there is no homework until the age of thirteen, and no exams until sixteen. Teachers are selected from among the best graduates, and everybody holds a government-funded PhD and earns a salary that's initially low but increases over the years.

because, even if it's culturally linked to youth, it potentially never stops.

Can one learn to play a musical instrument at the age of 60? And start to speak a new language at 70? The answer is always yes – although much depends on what has happened in the previous 60 or 70 years. The more a person has increased synapses and added myelin, even past their school years, the easier learning will be for them. Somebody who has practised sport all their lives can make a golf debut when they retire, but it will be more difficult for those who've never done any sport. Similarly, those who are accustomed to reading a lot [p225] will find it easier to learn statistics or Portuguese at 60 and over. But nothing is precluded.

ON	OFF
Knowing that you can learn to learn.	Believing that talent is predetermined.
Working hard is good: it means your brain is reorganising itself.	Letting yourself be scared off by fear.
Repetition is necessary: it's a rule of the memory game.	Forgetting that long-term memory forgets.
Varying interests (in order to find out more) and also routines.	Having a small number of interests.

7.4 IMAGINATION

What do an arrow, an anvil, an anchor, an astrolabe and an airplane have in common? They are all the fruit of human creativity, from the times of survival to the times of knowledge. And they are just a small part – starting with A – of a monumental river of creation that has been flowing for millennia. The thread that links poisoned arrows to intercontinental flights is called imagination.

With strong motivation, access to the widest possible

knowledge, and suitable attention given to it, imagination has had the evolutionary function of solving problems. How can you hunt that fierce animal without getting too close to it? How can metal be shaped? How can a boat be stopped from going adrift? How can you find your bearings in the open sea without land references? How can you cross the oceans so easily? And so forth, right down to the invention of zoos, zips, zesters, zithers and zoom lenses.

Wherever you may be at this moment, you have a clear perception of your environment and surrounding events. Let's do an experiment. Imagine that three Wild West cowboys suddenly appear in front of you, or three film celebrities, and try and imagine what will happen in the next 30 seconds.

Did the cowboys start shooting? Did Julia Roberts come and sit next to you? Whatever happened, your brain has drawn a parallel reality, generated by **divergent thinking**, or the choice from many possible alternatives. The perfect example of this process is that demonstrated by Albert Einstein, who, drawing on his breadth of knowledge, his deep curiosity about the unknown, and well-focused attention on the problems of the universe, discovered that time and space are different dimensions of the same spatiotemporal continuum.

But this is an inappropriate example, because it gives the sense that imagination is the exclusive preserve of Nobel Prize winners, when it is actually available to the entire human race. It's about drawing in your mind an alternative road to escape the traffic. It's writing a poem to flirt with someone. It's inventing a new fusion recipe. But it's also the door to creativity, usually defined as the production of original ideas that have an intrinsic value. Just like the arrow, the anchor and the anvil.

In post-industrial society, creativity has become classified as an essential economic resource. In 2011, the total sum of publishing, arts, design, fashion, film, music, TV shows and software represented roughly 3% of European GDP, 500 billion euros, and six million jobs. It has grown substantially since then. According to economists, creativity is destined to have an increasingly key role in global economic competition, because responses to modern challenges are shaped, at times even more than by the price, by the strength and novelty of ideas.

The so-called **knowledge economy** is a new industry that isn't – like so many industries before it – based on the strength of muscles or machines, but on that of divergent thinking. It's called 'knowledge' because it includes patents, trade secrets and various types of expertise, but we could also call it creativity economics because the starting and end points are valuable creations. It's hard to say when it began (long before the invention of the printing press?), but it has, for sure, a great deal of time before it to expand, change and increase its hold over the world now that digital communication has allowed us to cross geographical and temporal barriers. Just to be clear, the knowledge economy is the reason that, on Wall Street in September 2018, Alphabet Inc. (Google) was worth more than seven Fords, General Motors and Chryslers put together. From this early glimpse of the 21st century, it is clear that creativity is the most strategic economic resource there is. Which is why it is perfectly sensible to learn to cultivate it.

Every brain is equipped with an integrated system, made up of many different areas that become active whenever the user shifts his or her attention from the outside to the inside world. Discovered in 2001, the default mode network – the 'start-up' mode of brain mechanisms – is at the neural root of reflective thinking, be it in relation to oneself or to others, the memory of the past or the forecast of the future. To be precise, it is active when your mind drifts, when you daydream. And, consequently, also when you use your imagination to climb the high peaks of creativity.

The default mode network includes numerous and assorted areas of the brain, which have close axonic contact between them [p17]. As always happens in the most complex executive functions, the prefrontal cortex [p331] is significantly involved in the mechanism. However, parts of the cingulate cortex, which is located above the corpus callosum of the temporal and parietal lobes [p49], are also involved, as is the hippocampus [p45]. It must be mentioned that the default mode network also plays a part in self-awareness and consciousness [p126].

As people accustomed to creating and inventing know, that particular state of mind connected to abstraction and divergent thinking is reached through a peculiar kind of concentration that opens up the prairie of imagination. It's almost like a *click* that switches on the brain's creative mode. In her book *A Mind for Numbers,* Barbara Oakley, Professor of Engineering at Oakland University, calls it the **diffuse mode**. In a nutshell, the 'diffuse mode' consists of a 'diffuse' thought, capable of observing all the aspects of a problem, and is typically associated with the default mode network, as opposed to the 'focused mode', which is the rational and analytic attention of the prefrontal cortex. No brain, not even yours, is capable of activating both systems at the same time.

The methods currently adopted (including by multinational companies) for developing imagination are abundant and varied, because there isn't just one creativity model. Each of us can legitimately find our own formula, in tune with our own particular passions or inclinations; all we need is to know that it can truly fuel us. It may be strange to drag a 50-year-old manager to a creativity training session when he could have begun at a more appropriate time: his early school years. Children have that natural tendency to imagine, which they can either cultivate or abandon as they grow up, especially if encouraged or discouraged.

'Creativity is just connecting things,' Steve Jobs, the man who, through sheer force of thinking, created Apple, one of the most successful companies ever, once said. 'When you ask creative people how they did something, they feel a little guilty because they didn't really do it, they just saw something.'

You're bound to have met people who were more creative than you. But don't let that put you off. Imagination is an integral part of your brain equipment. Just imagine all the things you can do with it.

ON	OFF
Divergent and convergent thinking.	Convergent thought.
Using the motivation–attention–knowledge axis.	Using the demotivation–inattention–ignorance axis.
Believing that creativity can be increased.	Believing that creativity is reserved for those who already have it.
Learning to switch on 'creativity mode'.	Not even trying.

7.5 DECISION-MAKING

You will probably agree that every important decision is characterised by a rational calculation of alternative possibilities. Right? Well, not exactly.

Take Elliot. All in all, he was a happy, successful man, a good father and a good company manager. Until a tumour in one of his frontal lobes forced him to have an operation that ended up changing his inner world. It was as though he had become detached from everything, incapable of feeling even the slightest emotion. But the story of this anonymous patient, told by Portuguese neuroscientist Antonio Damasio in his book *Descartes' Error: Emotion, Reason and the Human Brain*, is even more dramatic. The interruption of his emotional circuit produced in Elliot a side effect nobody could have expected: instead of making him perfectly rational, it made him incapable of making any decision. The rest of his brain works perfectly well and is of above-average intelligence, as before. However, the apparent psychophysical normality, combined with the total inability to choose what to eat or with which pen to take a note, led him, within a brief period of time, to lose his wife and his job. Thanks to him, we now know that emotions don't get in the way of decisions – they are essential to making them.

This is not what people generally think. 'When we describe someone as "emotional",' Damasio remarks in the preface to the second edition of his book, 'it's usually a criticism that suggests that they lack good judgement. And the most logical and intelligent figures in popular culture are those who exert the greatest control over their emotions.' The conclusion of Damasio's book, however, is that rationality needs emotion. Hence, Descartes' 'error': it's not true that mind and body are dualistic and separate.

This is vital knowledge to have, if you reflect that decision-making is at the root of everyday life, social life and the planet's economic system. So-called neuroeconomics studies human behaviour in order to maximise decisions in buying products by weighing up the rationality and emotion of the consumer. And that's exactly what you should do too.

Procrastination, addictions [p193] and false memories [p192] are all manifestations of the irrational and impulsive side of the brain. Assuming that the 'sixth sense' does not exist (since there is nobody capable of making 100% right choices), intuition, i.e. the subliminal [p140] ability to make instant decisions even without having access to much information, is an extraordinary resource. Especially when a quick decision has to be made and there's no time for slow, reflective thinking.

Still, it is also true that 'gut feeling' can often be triggered by a cognitive bias [p189] dressed up as intuition, which could therefore turn out to be totally wrong. The first possible bias is to believe that, if intuition has worked once, then it will work again. Maybe it will. But it depends.

The executive functions of decision-making seem to reside in the anterior cingulate cortex (situated under the frontal lobes) [p49] in the orbitofrontal cortex (right behind the eyes) and in the ventromedial prefrontal cortex (even further back). But it's their deep connection with other parts of the primate, mammal and reptilian brain [p37], the cerebellum included, that establishes anatomically the coexistence between rationality, emotion and impulsiveness. As well as their oscillation through time.

Neither reason nor intuition always work impeccably well. They are influenced by emotions, the time of day, the food introduced into the system, the number of hours slept, the events experienced shortly beforehand, and current circumstances. For example, we recommend you avoid going to the supermarket at lunchtime: hunger would prompt you to buy food you would never eat.

Under normal conditions, the more important the decision, the better it is to mull it over. Partly to evaluate the situation better with more information and perfected reasoning, and partly to find the optimal physiological and psychological moment: usually after a night's sleep. This does not mean you need to be like Penelope weaving her cloth and postpone a decision ad infinitum.

The awareness of a possible cognitive bias (ranging from an empty stomach to faith in one's intuition) can be helpful. For example, if you agree that shopping properly when hungry is difficult, you can devise some countermoves: change your timetable, have something to eat or stick to the list of items written down four hours earlier on a full stomach. Rationality is anything but optional. You need it in everyday life. You just have to drop this whole idea of a *Homo oeconomicus* who makes perfectly rational choices – one of the foundations of classic economics – because, alas, it isn't true.

Decision-making, along with its twin, problem-solving, needs the backing of motivation, attention, knowledge and also imagination in order to function well. A decision can, in time, prove to have been wrong or a stroke of genius. It can turn out well or badly, and there's no need to blame reasoning or emotion. Mistakes help us to learn and start anew. This is why human beings can become truly wise as the years go by. As a matter of fact, wisdom is one of the last levers on your control panel.

ON	OFF
Knowing that decisions are influenced by emotions.	Believing you have everything rationally under control.
Choosing if and when to rely on intuition.	Relying totally on intuition.
If possible, choosing the right moment in order to choose better.	Postponing the moment of choosing in order not to choose.
Recognising your own cognitive biases and weighing them.	Bias? What bias?

7.6 COGNITIVE CONTROL

The marshmallow is not exactly an attractive-looking sweet. Manufactured industrially in the United States since the post-war boom, it's made from the *Althaea officinalis* root (also used for this purpose by Ancient Egyptians), sugar, egg and gelatine. It looks like a rubbery cylinder and is far less inviting than a chocolate. And yet it has been part of the history of psychology since the 1960s, when Stanford University decided to conduct a peculiar experiment.

Children between the ages of four and five are made to sit at a table with a marshmallow on it. 'You stay here,' the researcher tells the child, 'and I'll be back in fifteen minutes. If you manage not to eat the marshmallow, I'll give you another one once I'm back. OK?' And she leaves. The footage of little girls and boys staring adoringly at the sweet, trying to resist the temptation to gobble it down, is very entertaining. It has opened the eyes of the world to **delayed gratification**, the exquisitely human ability to give up on a dopaminergic reward [p25] in return for one that is bigger but delayed in time.

The Stanford researchers kept track of their subjects over time, looking for significant statistical correlations. Those who, as children, had resisted the appeal of the marshmallow (often with ingenious solutions, such as hiding under the table in order not to see it) reached a higher level of education as adults, as well as having a lower body mass. In other words, over the years they were able to exercise control over their temptation to not study, or to finish the whole chocolate cake.

Delayed gratification is one of the most important characteristics of cognitive control, the process by which behaviour is constantly readapted depending on aims and circumstances. Or, to put it the other way around, the process by which aims and personal projects influence behaviour.

Connected to the concepts of consciousness [p126] and free will [p145], cognitive control starts to develop as early as the age of four and increases until adolescence. At about sixteen, impulsiveness peaks, and after the age of twenty it consolidates and remains stable throughout adulthood. After the age of 70, it starts to decline. Important functions like attention [p157], operative memory [p64] and emotion management [p114] are largely dependent on cognitive control, the malfunction of which is linked to many neuropsychiatric disorders [p201].

Alongside delayed gratification, there is **inhibitory management**, or the brain's ability to restrain impulses because of a conflicting aim. The classic example is avoiding swearing at someone high up in the hierarchy in order to continue receiving your pay cheque at the end of the month. Habits and addictions, conversely, are examples of interrupted or faulty inhibitory management. So is the tendency to lose your temper or to allow yourself to be easily overwhelmed by emotions, be they good or bad.

Another important inhibitory function is the **suppression of irrelevant thoughts**. As you know very well, your brain can produce the strangest thoughts, from the comic to the macabre. You can let yourself be carried away by these hilarious or dramatic thoughts, but knowing how to say 'that's enough' and having the will to push them away is really very useful to the psychophysical wellbeing of a user like yourself.

States of anxiety are often connected to a vicious circle of thoughts that one cannot break. If a low degree of stress is helpful to some cognitive functions, severe and very prolonged stress has toxic effects on the organism and the brain. It must be avoided at all costs. Yes, but how is that **stress control** to be achieved?

This manual aims to summarise the brain, the most complex thing there is. But trying to summarise what occurs in the billions of neurons of over seven billion human beings would really be too much. Attaining a comprehensive level of cognitive control in that hotpot of action potential is even more complex than describing the brain itself. The wonderful diversity of the human race is revealed in a wave of actions and reactions, of hopes and disappointments, which oscillates between highs and lows over time.

Discoveries such as neuroplasticity [p69], insights such as

those into the *growth mindset* [p72], and experimental results such as those in positive psychology [p123] collectively reverse the old fatalistic view of a static, immutable brain. There is no prescribed fate; we are not the slaves of our character, nor of our circumstances. Once that has been understood, every brain must use the levers of its own cognitive control.

It's true that it is good to train delayed gratification from an early age (as Asha Phillips's book *Saying No: Why It's Important for You and Your Child* implies), but you can also learn it as an adult. Inhibitory management is not just useful to avoid being fired, but for the entire range of social and relational activities. Society, for example, does not like brains that lose patience easily. Irrelevant thoughts can impair learning, studies and work; if, in addition, they create anxiety, they can contribute to depression. Usually, this kind of control is learned at the youth stage of brain development, but, with a little effort and exercise, it can also be learned as an adult. In addition, being able to recognise chronic stress [p196] and doing whatever it takes to relieve it contributes to the integrity of the general cognitive system.

Forty years later, those children who had not resisted the attraction of the marshmallows were fatter and less educated than their peers. But over those forty years, they could have changed path and direction, if only they had known how.

So there you have it: your brain doesn't have any more excuses.

ON	OFF
Learning to delay gratification.	Better an egg today than a hen tomorrow.
Learning to control impulses.	'You are a total b******!'
Learning to suppress irrelevant or anxiety-causing thoughts.	The painful pleasure of dwelling on things.
Learning to keep chronic stress under control.	Stress a go-go.

8.0 MODELS

BRAINS ARE MANUFACTURED IN TWO POSSIBLE versions. The F Model° (female) is the standard one and the M Model° (male) requires a series of adjustments during the construction phase. It is not possible to pre-order the desired version. The reason for this is the assembly method, triggered by a phenomenon we could call random. The joining of a female ovum with a single sperm unites half the genetic heritage of the mother and half the genetic heritage of the father. From the mother's side, the 23rd pair of chromosomes (the one dedicated to gender) is composed of a double XX, which means that the ovule will always carry an X. From the father's side, on the other hand, the chromosomic pair is XY, and the sperm can contribute either of the two. If the race to conception is won by a sperm with a Y, the new brain will be produced in a male version. If one with an X wins, the brain will be female. Anne Boleyn, who was beheaded by Henry VIII for not giving him male heirs, deserves an apology at the very least.

During the first eight weeks of assembly (also known as 'gestation') all brains are identically feminine. From that point on, however, a discharge of testosterone in M brains triggers a sequence of small but radical structural reforms which, after another 30 weeks of assembly, complete a brand-new M brain, linked to a perfectly functioning body and finally able to live its own life.

But it doesn't end there. During the whole assembly process, the brain is connected directly to the mother's biological factory, with which it shares blood, nutrients and hormones. The last of these inputs influences construction of the brain with differing characteristics which we might have deemed typical of the F Model° or the M Model°, resulting in something of a mixed deck. This is probably what F brains refer to when they talk of their 'masculine side' and M brains of their 'feminine side'. It is now an established fact that genes, hormones and brain structure contribute to sexual orientation, which can be heterosexual, homosexual, bisexual or even asexual. From studies on twins to fMRI scans, it is clear that every brain shapes the orientation by itself: a homosexual brain tends to have more characteristics that are similar to the brains of the opposite sex. It's no coincidence that the gruesome old custom of coercing a person to change sexual orientation has caused huge suffering but no 'success'.

For all these reasons, some claim that there aren't really two very different versions but just one with cross-over characteristics. The brain is the brain. However, the operative function of each model is visibly different and it's interesting to compare them, partly thanks to their inherent supply of humour.

8.1 COMPARING THE F MODEL®
AND THE M MODEL®

Nobody expects a lioness to act like a lion. Nor a hen like a cockerel. So you can picture just how complicated it can all get when it comes to a man and a woman. The behaviour of the human species is so dimorphic (from the ancient Greek *dimorphos*, 'which has two shapes') that for a long time it was believed that brain architecture was, too. The paradox is that it barely is at all. For every study that suggests a divergence between the two models, there is another that shows the opposite. Differences do exist, certainly, but they are not as dramatic as respective behaviours would lead us to believe.

In the following tables, we give you the established views on the dimorphism of the human species, which should not be interpreted as absolutes, but rather as prevalent factors in statistical distributions, where the difference between the two models is sometimes trivial, to say the least.

F MODEL® (XX)	M MODEL® (XY)
The X chromosome contains about 1,500 genes that encode essential proteins, including for brain development. Having two Xs, the F® possesses a back-up copy.	The Y chromosome (not by chance considered a 'gene desert') contains less than 200 genes and only 72 of these encode proteins. The M Model® has only one X, with no back-up.
A more efficient brain (it uses proportionally less glucose).	A 10% larger brain on average (in proportion to the body).
A thicker cerebral cortex, a larger thalamus.	Larger amygdalae, hippocampus, striatum and putamen.
A structurally more complex corpus callosum.	A structurally larger corpus callosum.
More interhemispheric connections (between hemispheres), which, some claim, facilitate communication between analytical and intuitive thinking.	More intrahemispheric connections (within the hemispheres) which, some claim, facilitate communication between perception and action.

Contrary to what has long been believed, the two cognitive systems are not significantly different. Neither are their levels of intelligence: the F Models° were slightly below the M average in intelligence tests only when they *felt* inferior to males. Once the prejudice was removed, no differences were recorded.

Behaviours, on the other hand, are blatantly different and start to diverge from a young age. But here we have the usual question: is nature or nurture more influential? Do girls prefer dolls and boys lorries because it's determined by their genes or because they both learn to behave according to the usual path of imitation and reward? Everything leads us to believe that nature prevails, but that nurture has a strong impact, too.

F MODEL® (XX)	M MODEL® (XY)
Prevalence of language abilities (talks a lot).	Prevalence of mathematical abilities (talks less).
Beats the M Model® in the perception of other people's emotions (empathy, social relationships).	Beats the F Model® in spatiotemporal navigation (orientation).
Stronger emotional experiences, a more solid emotional memory.	Overestimates his abilities.
Stress (at exams, for example) lowers performance.	A certain amount of stress tends to heighten performance.
Behaviour control.	Tendency towards risk taking.
Maintains visual contact and face-to-face communication with female friends.	Doesn't look friends in the eye and remains in lateral or angled position.
Recent social mutations have added the possibility of being managing directors, joining the military and being sexually more enterprising.	Recent social mutations have added the possibility of taking paternity leave, crying at the cinema and using cosmetics.

The male peacock, with its multicolour fan, is much more flamboyant (and clumsier) than the female peacock. The male mandrill is three times as large as his mate. This dimorphism, like all examples of dimorphism, has a specific evolutionary function: to get to the bedroom.

It's no wonder that, when talking about sex and its ancestral influences, the 'prevalences' in the two models end up looking increasingly like stereotypes from women's magazines

('Ten effective tips for getting him back') and also men's magazines ('Seven things you can tell her to drive her wild in bed').

F MODEL® (XX)	M MODEL® (XY)
Thinks about sex in moderation, except during ovulation when (possibly at the subliminal level) she is more proactive.	Thinks about sex with no moderation, several times a day, every day. If he doesn't admit to it, then he's lying.
Considers sex as a means (evolutionarily, a stable relationship is essential to the survival of offspring).	Considers sex as an end (evolutionarily, spreading genes is essential to the continuation of the species).
In choosing a partner, status is more important than physical appearance.	In choosing a partner, physical appearance is more important than status.
The higher the self-esteem, the lower the tendency towards promiscuity.	The higher the self-esteem, the higher the tendency towards promiscuity.
In case of jealousy, considers 'emotional betrayal' more serious (it puts the relationship at risk).	In case of jealousy, considers 'physical betrayal' more serious (it puts the certainty of paternity at risk).
During the love phase, prevalence of dopamine, oestrogen and oxytocin.	During the love phase, prevalence of dopamine, testosterone and vasopressin.
'Why do women need an orgasm?' many (M brains) have asked in the past. In actual fact, the female sexual experience is apparently superior and more varied in quality.	Believed (or believes?) that his orgasm is the centre of the universe.
Can be psychologically induced to simulate orgasm.	Can be psycho-hydraulically forced to give up the ghost.

The differences between the two genomes (starting with the X and Y chromosomes) and between their respective hormonal systems also contribute to exposing the two brain models to different prevalent pathologies. The futuristic dream of tailor-made medicine can only come true when medical science is finally able to produce and prescribe treatments that are specific to each of the models.

F MODEL® (XX)	M MODEL® (XY)
Depression, anxiety.	Autism, schizophrenia.
Shopping (some say she has a tendency towards gambling but in actual fact she's far less predisposed towards it than the other model).	Alcohol, drugs, gambling.
Pre-menstrual tension carries 200 *possible* different physical and emotional symptoms that *can* last 6 days and temporarily alter the view of the world.	Structural inability to understand that the pre-menstrual syndrome can regularly reoccur in F Models® every 28 days.
Is not affected by haemophilia, Duchenne muscular dystrophy and (almost never) colour blindness.	Is not affected by Rett Syndrome.
The 'gentle sex' has a higher pain threshold.	The 'strong sex' could not tolerate the pain of childbirth.

The list of differences between the two models could go on and on. For the sake of brevity, we'll add only four more.

F MODEL® (XX)	M MODEL® (XY)
95% of ultra-centenarians are female.	5% of ultra-centenarians are male.
Started voting much later.	Earns more on average.
Feminism is politically correct.	Male chauvinism is politically incorrect.
Always wonders: 'Why doesn't a man think like a woman?'	Always wonders: 'Why doesn't a woman think like a man?'

9.0 FREQUENT PROBLEMS

DRAPETOMANIA IS A MENTAL ILLNESS WITH terrible consequences. Discovered in the mid-nineteenth century by American surgeon Samuel Cartwright – he wrote a treatise about
it – it was considered an inexplicable personality disorder: it
would prompt slaves to run away.

Although, two centuries later, we can laugh about this totally
made-up illness, it's harder to do so about other imaginary
illnesses, such as homosexuality, which sadly featured on the list
of mental disorders as recently as 1973.* Throughout the centuries, every 'anomaly', from chronic stress to the destructiveness
of Alzheimer's, has always had one stigma or another attached
to it. You could be the village idiot, the lunatic to be locked up
or the witch for burning. Nowadays, you could be excluded
because you're chronically depressed, or pitied because you're
autistic.

No, the brain is definitely not perfect. Evolution has scram-
bled, duplicated and added structure upon structure, at times
collecting imperfections. And there's more. The genetic information written in every neuron may have predisposed you
to a mental illness. Predisposed does not mean predestined:

---

*   In 1973 the American Psychiatric Association voted by a majority to delete
     homosexuality from the *DSM (Diagnostic and Statistical Manual of Mental
     Disorders*, the 'Bible' of mental disorders), but it was removed only in 1987.
     On the other hand, the *ICD (International Classification of Diseases)*, the
     classification of all diseases, not just mental, edited by the World Health
     Organization and used in the rest of the world, didn't remove it until 1992.

studies on schizophrenia in identical twins (who have 100% identical genomes) show that the probability of the other twin being affected is only 50%. Genes play a gigantic role but are not destiny. The brain can be the victim of trauma, which, depending on the brain area concerned, can produce unpredicatable effects, including a change in personality. It can be the victim of such severe psychological trauma that vision of oneself and the world might be altered, or – and there's no explanation for this either – the trauma might not make the slightest dent.

There are no two identical depressions in this world. There is no phobia, addiction or disorder that produces the identically same effects in two different human beings. Perhaps not even in the case of neurodegenerative syndromes, which follow more or less the course described by Dr Parkinson (1817) and Dr Alzheimer (1906), do two exactly identical clinical cases exist. Not to mention the case of autism, where the expression **autistic spectrum disorder** has been adopted, precisely because it's not a single colour but an entire palette.

One can suffer from chronic depression and have a pretty much normal life, or be totally debilitated by it. One can have schizophrenia and not hear non-existent voices. One can have a mild gambling or shopping addiction, or else have a totally destructive form of it. As though this weren't enough, the dividing line between mental illness and 'normality' is so uncertain and blurred that normality requires inverted commas. What would you call 'normal'?

If we take the word of British research, which estimates that one person in four has a mental illness (however mild), out of the seven and a half billion brains lurking around the planet, about 1,800 million – that is all the citizens of Europe and China combined – have a problem. Except that we don't take its word as gospel.

It's too difficult to define the categories and who belongs to them; it's foolish to apply to the whole world the average of just one country and not consider that some disorders are more prevalent in the West than in Asia or Africa. But one thing is undeniable: problems of a cerebral and mental nature are more common than you would think.

We present here a selection of categories (far from a complete list) only in order to inform you of the most common problems experienced by brains like yours. It's divided into two parts, if nothing else to keep a brain's **computational errors** – such as the almost always harmless case of synaesthesia – separate from actual **malfunctions** that lead to a myriad of possible disorders and require the intervention of specialised professionals.

9.1 PROCESSING ERRORS

Distraction is a banal calculation error but it can be fatal on a motorway. Optical illusions are a calculation error but are certainly less insidious than hallucinations. The illusion of something that doesn't exist is also a calculation error, such as that of not having a limb or living in a world where everybody is always secretly conspiring against you. Personality extremes could also be listed among calculation errors, for example narcissism ('I am a god') and nihilism ('I am nothing'), or psychopathy, an 'antisocial personality disorder' characterised by scarce empathy, plentiful egocentricity and zero remorse.

There are those who are not afraid of practising free-climbing and those who wear masks because they are terrified of germs. Those who live just for the sake of making merry on a Saturday night, flitting from one party to another, and those who are affected by anthropophobia, the 'fear of people', an extreme and pathological form of shyness. Those who love travelling by plane and long to accumulate the famous ten million air miles (like George Clooney in the film *Up in the Air*) and those so terrified of flying that they cry and tremble for the entire intercontinental flight. Within the immense variability of the *sapiens* brain, an immense variability of calculation errors is concealed. Not all are 'disorders', not all are destiny, but in some cases they can be terrible. Here they are in (alleged) order of gravity.

9.1.1 Synaesthesia

What did Franz Liszt, Wassily Kandinsky and Duke Ellington have in common? All three were synaesthetes. In their brains, the perceptions of one sensory channel (hearing sounds) would trigger another sense (the vision of colours).

There are as many variations on the theme of synaesthesia – from the Ancient Greek *syn*, 'union' and *aisthànesthai* 'perceive', 'to perceive together' – as there are variations by Liszt. There are those who feel touched when they see somebody else being touched. Those who experience taste when they hear certain words being uttered, such as 'adventure' tasting a little like raspberries. Those who associate letters, numbers, names of the days, weeks and months with anthropomorphic identities, such as Thursdays being male, overweight and irritable. Those who have auditory-tactile synaesthesia, i.e. feel physical signals in response to sounds. And so forth, with tens of other possible sensory intertwinings. Thanks to YouTube, many people have discovered that they have an ability that could be included within the borders of synaesthesia. It's called ASMR (Autonomous Sensory Meridian Response) and produces a strange but pleasant physical sensation at the back of the neck in response to two possible phenomena: listening to a whispering voice or to light friction sounds; or else – strange as it may seem – seeing *someone else* doing precision work with their hands.

Synaesthesia is mostly pleasant, and apparently helpful to artistic output. At times, however, it can be torture. This is the case in the variant called **misophonia**: in response to specific sounds or noises, one experiences fear, hatred and disgust.

According to some authorities, synaesthesia could derive from a lack of 'pruning' of some neural connections during childhood [p79], resulting in a few sensory pathways that talk too much to one another.

9.1.2 Placebo and Nocebo

The brain is able to believe it – and even to lift itself out of the doldrums – when it tells itself that things aren't so bad after all [p129]. But is it really so easy to cheat the central nervous system? Is it possible that the centre of intelligence, which is also the centre of doubt and suspicion, can so easily be taken for a ride? If you happen to have a professional crook or a surgeon among your circle of friends, try asking them. Both will answer yes. But the most surprising stories will come from the doctor.

It has been known for centuries that the symptoms of a disease can be eased by making the patient believe that they are being treated, but it was only in the 1700s that this bizarre psychological effect was christened *placebo*, from the Latin 'I will please'. This phenomenon, tested and re-tested with bogus medicines and even simulated surgeries, is essentially a mystery. We know that it can involve neurotransmitters and activate different areas of the brain, from the strategic prefrontal cortex [p53] to the emotional amygdala [p44]. We know, too, that this doesn't work on all patients but only some. It is suspected that the difference between the two groups has genetic roots, but there is no conclusive proof of that.

This kind of psychological trick that *cheats the brain into contentment* – not only with a pill but with an entire ceremony officiated by white coats – helps relieve the symptoms of an illness but seldom cures it. It can also produce the opposite effect: cheating the brain *into discontent*. Among patients convinced that their medicines have a negative effect on their health, some will truly start to feel unwell. Another example of a calculation error the brain can make. They call it *nocebo*, 'I will harm'. It is the peculiar flip side of a strange coin.

9.1.3 Cognitive Biases

Economic theory sees the human being as a totally rational agent interested in maximising his or her own profit, the so-called *Homo oeconomicus*. This concept of rationality is somewhat unfounded, since the two thought mechanisms – above and below consciousness [pp127, 140] – can mix things up. This means that even in the most rational convictions, thoughts and behaviours, you could still potentially be the victim of a long series of subliminal cognitive biases. This throws doubt on rationality theory. It's no wonder that a new interdisciplinary field has emerged, neuroeconomy, which studies the iceberg of the decision-making process, where the 'visible', conscious part (the one that apparently selects the alternatives from which to choose) is only the tiny peak that emerges from the water.

Here are a dozen of these cognitive calculation errors (only a small selection of those described by psychologists). You have probably already come across a few of these in your life.

Apophenia. Just as the visual system specialises in finding patterns in everything that comes from the retinae [p102], the cortex deduces the presence of recurring elements in totally random events, for example in lottery draws or any kind of divination, from tea leaves to tarot cards.

Gambler's fallacy. Similar to apophenia. It consists in believing that, since 'heads' came out five times in a row, the next time one tosses the coin, it will probably land on 'tails'. Mathematics disagrees: the probability is 50-50 with every toss.

The bandwagon effect. The tendency to believe in something because many other people believe it. All the most unpleasant cases of mass folly recorded throughout history have included this common calculation error.

Hindsight bias. Past events that suddenly appear foreseeable: 'I knew it,' you think. It's absurd, and yet all those who buy and sell shares know this effect well, and are reluctant to abandon it.

Confirmation bias. Any new piece of information – whether true or false – confirms existing beliefs and obviously refutes opposing ones. It's more frequent in consolidated beliefs, such as faith in religion, politics and sports.

Declinism. The distinct sense that everything is worse than before, spiralling into pessimism. Naturally, this can happen in life. But that it should be applied to *everything* and *always* is rather improbable.

Anchoring. The first piece of information perceived becomes the anchor for a subsequent thought process. A trick used by the seller who first calls out the (high) price of a second-hand car, after which any other model and a lower price seems like a real bargain.

Conservatism. When something new is viewed with suspicion and its value under-estimated in comparison to that of previous convictions.

Novelty bias. All new information, be it bizarre, entertaining or with a strong visual impact, takes priority in cognitive mechanisms, while expected and 'normal' information is put on the back burner. At the end of the TV news, the steel workers' strike makes less of an impression than the guy who threw a custard pie in the Queen's face.

Stereotyping. Since the memory system is based on associations and categorisations, when the brain possesses only partial information, it completes it automatically by resorting to associated categories [p192]. *Voilà*, the stereotype.

Illusion of transparency. It's true that, thanks to the empathy mechanism [p132], a brain can perceive another person's mental state. But there's a huge difference between this faculty and really *knowing* what they think. If you think you know what someone else thinks, then we are sorry to inform you that it's an illusion.

Bias blind spot. If you notice that all these biases influence your friends', colleagues' and relatives' thinking far more than yours, then be aware that this is a bias.

9.1.4 False Memories

'Intelligence is your wife, imagination your mistress and memory your maid.' Victor Hugo's quip, very nineteenth century and somewhat politically incorrect, views intelligence as an asset worth holding on to, imagination as a fling, and memory as service you're entitled to. Shame this service isn't all that reliable.

Memory is reconstructive, not reproductive. Put simply, it's not like a video recorder that reproduces film frames, but rather like a warehouseman who has to reconstruct all the pieces of an event, interlinked by a chain of mental associations. Whenever it is recalled, every memory may be slightly wrong, and further errors may be added the next time. In some cases, it may become wholly unreliable.

A host of experiments and psychological studies have proved without a shadow of a doubt that memories are weak, that they deteriorate, and that they can be altered from outside and even 'implanted' quite easily from nothing. This raises three big issues. First, Elizabeth Loftus, a famous scholar of false memories, claims that in the US over 300 prisoners have been set free after decades thanks to DNA evidence: three quarters of these had got into trouble because of at least one witness with a defective memory. Secondly, 'fake news' circulated through the Internet has become rife since 2016, facilitated cerebrally by a collection of 'hearsay' memories and a plethora of cognitive biases. Thirdly, it's a very useful asset for leaders of totalitarian regimes used to implanting fake memories in their people, as in the case of 21st century North Korea.

On the pathological gravity scale, there's also a false memory syndrome that carries with it all the symptoms of a traumatic experience, except for the fact that it is totally imaginary.

According to Dr Loftus's controversial view, it is often associated with therapies based on the recovery of past memories.

9.1.5 Habits and Addictions

Quite an invention, habit. It helps one to drive a car without having to learn how to do it all over again every time. It helps one to avoid tooth decay because, no sooner is it installed than the need to brush one's teeth arises automatically. Ultimately, it can also help one live longer, since one can develop the habits of physical exercise, drinking a lot of water and keeping away from trouble.

In preparing the habit module, several million years ago, evolution arranged it into the structures and functions of three existing systems: learning (conditioning in particular) [p60], memory (association mechanism) [p64], and reward (the dopaminergic spring of motivation) [p141]. Naturally, everything has evolved over many centuries to give you an integrated, speedy service, compatible with your *sapiens* system version [p10].

Quite a curse, habit. In some cases, it prompts people to eat even without the stimulus of hunger, whenever they watch television. It forces people to smoke the cigarette they don't feel like immediately after a cup of coffee. It requires the compulsive purchase of unnecessary items whenever the mood is low. It's automatic – apparently conscious and yet totally unconscious.

Habit gets anchored more or less rapidly but always in an incremental way, finally producing a kind of classic Pavlovian conditioning [p60]. At that point, all it takes is a specific go signal associated with something else – TV, coffee, emotional state – for a burning desire to obtain a reward to be triggered: a slice of cake, nicotine, or the umpteenth pair of shoes to be abandoned at the bottom of the wardrobe. When this desire becomes irrepressible, obsessive or indispensable, habits really become a curse. It's called addiction.

There is addiction to exogenous substances (food, alcohol, nicotine and various drugs), which tickles the reward system, usually through pre-installed specialised receptors in the brain, such as cannabinoid and opiate receptors. In the case of the latter, addiction can be devastating because tolerance (the need to increase the dose), abstinence and relapse are very powerful. Few substances create as powerful an addiction as heroin and cocaine. Those who quit smoking feel the *physical* need for nicotine but withdrawal symptoms don't last more than five or six days. What makes it difficult to abandon cigarettes is hidden beneath the threshold of consciousness.

But there is also addiction to specific behaviours, not all of them expected at the outset. Over the past 10,000 years, at a snail's pace to begin with, then at the sustained pace of present times, technological acceleration has been incredible: occurring over a period of time too brief to allow evolution to keep up. Consequently, shopping, television, video games, porn and gambling can on occasion get the upper hand over rationality. And with a non-negligible new detail: since the early 21st century, these are all accessible through the planet's digital network, with the result that compulsive shopping, auto-eroticism and hands of poker are available 24 hours a day, 365 days a year. And far from prying eyes.

Not all brains are equally inclined to turn any dopaminergic piece of entertainment into an impossible-to-resist obsession. Some people aren't remotely tempted by the possibility of overdoing it with snacks at any time of day, or with cigarettes or television. Many develop unwelcome habits here and there (nail biting, using Facebook too much), bad behavioural traits (always getting angry, seeing only the negative side of things, doing no physical exercise) and perhaps even develop one or more addictions that are sometimes unmentionable. But there are some that can really lose their heads.

Just take a short visit to Las Vegas and you'll see how gambling, nicotine, alcohol and sex go cheerfully hand in hand. There is also a genetic link between different addictions in their most acute forms. In these cases, it's advisable to seek the assistance of health institutions and voluntary organisations that deal with specific problems. Particularly when the user considers it to be serious.

In spite of individual differences, the mechanism by which habits and addictions are formed is always more or less the same. Before going to bed or on waking up in the morning, you promise to yourself not to repeat this action, both yearned for and unwanted at the same time: to stop gulping down the drink that shortens life, overplaying the video game that wastes time, gorging on the snack that increases the waist-line. It seems like the rational thing to do, right? And yet a little later, breaking the promise to yourself becomes totally rational: at least from the point of view of placating the reward system *immediately*. The self-improvement project is post-poned until tomorrow. And the pattern is repeated.

The reward system definitely includes a small conscious and deliberate side (such as the pleasure of doing one's job well), but it is dominated by the ancestral part of the brain, which prefers short-term pleasure. Being able to postpone pleasure is a very useful function throughout a whole lifetime. It's precisely for this reason that the brain cannot be installed correctly if, during the first few years of life, nobody ever says no to it. It gets used to a fantasy world that doesn't exist [p174].

Fortunately, being able to postpone pleasure is a habit implantable into any system. Similarly, we are truly happy to inform you that by cultivating positive counter-habits unwanted habits can be uninstalled, including those that seem uncontrollable. With this trick, the rational brain can, if not take control, then at least alter and influence the automatic one.

American journalist Charles Duhigg summarises it beautifully in his book *The Power of Habit*. Let's take a brain with the bad habit of eating a slice of cake in a coffee shop every day after the 3 p.m. meeting, even though cholesterol tests suggest that, rationally speaking, it shouldn't. The end of the meeting is the **signal**, attacking the cake is the **routine** and the consequent effluvium of endorphins, dopamine and sugars is the **reward**. Excuse the banal example – it's up to you to adapt it to the habits most relevant to you. In the final analysis, the prescription consists of identifying the signal that triggers the need for a substance or behaviour, and at that signal, replacing the routine habit with another habit still able to provide a reward, however small.

After the 3 p.m. meeting, you could go and have a chat with your colleagues (social activities produce dopamine) and drink a nice glass of water (while projecting into the future a more slender and 'sexier' image of yourself). By repeating this pattern – not as enjoyable at first – you can soon erase the unwanted habit. With a little more effort, even an addiction.

Of course, it's easier said than done. Even so, it's good to have the basic idea clear in your mind: the automatic vicious circle *can* be broken by a virtuous circle using the same mechanism. Your brain is plastic, don't forget that.

9.1.6 Chronic Stress

The 'stresses and strains of modern life' began several million years ago. That is when evolution gradually introduced to planet Earth an extraordinary automatic security system called fear [p621]. It's likely that the (not exactly modern) life of a reptile 300 million years ago would have been slightly more stressful than the one you're leading nowadays. If you think about it, only a few centuries ago, human life – in the absence of fair laws, supermarkets, birth control and antibiotics – must have been rather stressful too.

As in the case of fear, the hypothalamus [p46] wastes no time in responding to stress: it orders the adrenal glands to *instantly* produce adrenaline [p24]. This hormone, which prepares for fight or flight, raises blood pressure and heartbeat in order to increase blood flow to the muscles, so you can either hit someone or run. Nowadays, many brains (but not all, of course) find this such a pleasant sensation that they happily pay to see a horror film or to parachute down from a cliff.

If the alarm is continuous, however, the adrenal glands have another string to their bow: cortisol, often called the 'stress hormone' [p27]. The difference between stress and fear is played out entirely in the fourth dimension, time [p111]. Fear was created so that we could avoid becoming a predator's lunch and would last as long as necessary for us to survive: a matter of a few minutes. Stress results when a strong state of anxiety goes on for months or years, as a result of the death of a loved one, for instance, or the unhappy conclusion to a marriage, or perhaps an exhausting job in a hostile environment. Just pick your favourite stress – as long as it's prolonged in time, that's where cortisol levels can overflow.

Cortisol inhibits the immune system, interferes with the endocrine system and attacks the hippocampi in particular [p323], in the worst-case scenario damaging them physically. That's why the stress hormone interferes with memory and learning mechanisms, which are regulated by the hippocampi themselves (and that's why it's not a good idea to customarily wield threats and punishments in a classroom) [p162].

Stress is therefore a bad thing, right? Wrong. If it didn't exist, no athlete would ever be able to push themselves to the next level. 'When you have the ball at your feet and tens of thousands of voices egg you on to run to the goal,' footballer Roberto Baggio once said, 'adrenaline gives you wings.' A certain amount of stress can be recreational. It can even be creative, in the sense that it maintains that state of mental alertness necessary to write a book by the deadline set by the publisher. Many studies confirm that, in the right dose, stress increases productivity in the workplace. As long as it's not raised above a certain level (and some 'bosses' do *that* very well), in which case productivity decreases. It's all a matter of measure.

On occasion, the stress measure can be not just full but overflowing. That's the case with **post-traumatic stress**, which produces the aforementioned effects, only multiplied ten or a hundred times, and can include permanent damage to memory. It happens in one-off cases of rape, violence or abuse at an early age. But it can also be generated on an industrial scale. The US Department of Veterans Affairs has declared officially that the number of post-traumatic stress cases after the Vietnam War was 830,000. Nothing for the Pentagon to be proud of.

Stress is not in itself a calculation error of your central nervous system. Let's just say that things don't quite add up when the fear mechanisms are activated too strongly and for too long. That's why stress management is an important if not essential part of cognitive control [p174].

According to the World Health Organization (WHO), stress and other mental disorders are more prevalent in Europe and North America compared to the rest of the world. In other words, places where the capitalist system and 'modernity' originated. It's partly the fault of modern life if long-established stress mechanisms puts a strain on the brain.

9.1.7 Phobias and Illusions

According to WHO statistics, fear of the future is more widespread in Western society than in the rest of the world. This type of fear is also known as anxiety.

When all is said and done, anxiety consists in worrying about events that are yet to take place, however unrealistic, improbable or impossible they might be. As usual, the brain responds with fear mechanisms: a hormonal release that raises the heartbeat. Just do a little experiment on yourself. Concentrate. Close your eyes and start drawing in your mind an event you fear dreadfully, and go on picturing details and consequences for a minute. You will feel the cardiovascular system reacting in your chest. And yet a minute later, once the exercise is over, everything goes back to normal. If, on the contrary, the negative thought starts going round and round in circles and is never interrupted by a positive counter-thought, then it's reinforced by the memory's automatic mechanisms [p64] and the state of anxiety becomes chronic, with myriad variations, some very debilitating.

In extreme cases, it can become a phobia. In other words, persistent fear of a situation, which in some cases triggers panic attacks, characterised by the vicious circle that goes into a spin without anything virtuous to ease it. Even if we put aside the best-known phobias (spiders, snakes, large spaces, small spaces, public speaking, flying and, obviously, death), the number of those officially listed is truly high. One can have an irrepressible fear of passing time, demons, dentists, cold, sun, the colour red, germs, numbers, smells, dreams, mirrors, sex organs, being looked at, being alone, and so forth.

There is nothing new: the neural substrates of fear are always the same and involve the amygdalae [p44] as well as the hypothalamus-pituitary-adrenal gland pathway [p46]. Phobias usually spring from psychological traumas, but not without some genetic help. It's been proved that, in many cases, cognitive behavioural therapies manage to alleviate, if not actually erase, many phobias. There is footage, available on YouTube, of former arachnophobes quite happily holding large hairy spiders in their hands.

At the most severe end of the scale of such disorders, though not necessarily connected to fear, we find a range of mental miscalculations people can suffer based around a single fixation – so-called monothematic delusions. These can derive from traumas, brain damage or more serious mental illnesses. Some believe that a friend or relative has been replaced by a double. Others think that when they meet several different people, they are in reality just meeting one person capable of taking on different identities. Others don't believe that they are the person reflected in the mirror. And others, after a stroke, categorically rule out the idea that their left arm, say, or the entire right side of their body belongs to them. This last case is known as somatoparaphrenia.

We could go on. A brain affected with **body integrity identity disorder** longs to lose a specific limb through amputation, often with a sexual excitement component, while with **acrotomophilia**, the burning desire is to have sex with an amputee.

Here we are on the brink of extremes, where calculation errors become malfunctions.

9.2 MALFUNCTIONS

In the final stages in the evolution of *Homo sapiens*, a few drastic changes in genetic expression have contributed to the increase in brain activity, and to the quality and plasticity of synaptic connections. There are those who believe that this increase in complexity has left the brain vulnerable to **neuropsychiatric** and **neurodegenerative** conditions such as schizophrenia and Parkinson's disease respectively. Although other animal species are not immune from them, humans seem to be markedly more vulnerable to brain malfunctions.

9.2.1 Autism

Derek Paravicini lives in London and has been playing the piano since the age of two. Now that he is nearly 40, he appears in shows and TV programmes where the audience makes requests and he plays any song from memory. Apparently he knows over 20,000 tunes. Born prematurely and the victim of the wrong therapy in the incubator, Derek is blind and suffers from problems caused by abnormal brain development during childhood. If he had been born a century or two earlier, he would have been hired by a circus or a freak show. Had he been born even earlier, he might have been locked up or even terminated.

Derek is autistic, like 25 million other people in the world. Now that disorders of the autistic spectrum are no longer labelled the work of the devil, they are mostly treated and respected. No two cases are the same. To generalise, we might say that people with autism experience, to a very variable degree, difficulties in interpersonal relations and communication, restricted (if not obsessive) interests and a preference for repetitive behaviours. In some cases, the impact of the disorder is minimal. In others, it's considerable. Some autistic people sometimes become *idiots savants,* like Derek Paravicini, or Raymond, the character played by Dustin Hoffman in the film *Rain Man*.

Other disorders on the autistic spectrum include neuro-development conditions such as **Asperger's Syndrome** (which affects the ability to understand others, not language or intelligence) and the 'pervasive developmental disorder not otherwise specified' (an atypical form of autism because it appears only in adulthood, and doesn't yet have a proper name).

Despite a deeply rooted myth, there is not the slightest evidence that autism is linked to childhood vaccines. Its causes are unknown but genetic predisposition is proved by its prevalence in identical twins. It is most common in M° Model brains [p178].

9.2.2 Chronic Depression

Defining depression is an arduous task for two reasons. Firstly, because the term is used inappropriately ('we have lost the World Cup, so I'm depressed') as well as in serious clinical cases. Secondly, because a mood that tends towards being low can be caused by a malfunction of the central nervous system, but also by a myriad of other conditions and disorders: auto-immune diseases, bacterial or viral infections, eating disorders, endocrine system disorders, concussion, multiple sclerosis, tumours or other mental illnesses (such as **bipolar disorder**, where depressive phases alternate with manic phases), which require different treatment and care.

Leaving aside sad but temporary states of mind that every

brain may experience in life (but which are not classified as malfunctions), chronic depression is a phenomenon of great importance. According to the WHO, about 300 million people are affected by it in the world, and this number is destined to increase significantly by the middle of the century. Sadness, anxiety, lack of hope, a sense of void and pointlessness, also in some cases a sense of guilt. These are all symptoms that lead to a withdrawal from social activities and a general loss of interest. This is caused by a combination of genetic, biological, environmental and psychological factors: a tragic event can trigger a domino effect. The US National Institute of Mental Health says it is 'clinical depression if it's there almost all day long, and almost every day for at least two weeks'. Two weeks may not be very long but if the phenomenon is prolonged and the depression is moderate or severe, then it's necessary to seek the help of a specialist.

In the last 20 years, more and more people are finally coming round to the idea that depression is caused by a chemical or biological imbalance in the brain, after centuries of incomprehension and denigration of those affected. It's no coincidence that this new attitude has been encouraged by marketing on the part of pharmaceutical companies, which advertise SSRI drugs on television (US law allows it) with the same nonchalance as they would with cough mixture.

Selective serotonin reuptake inhibitor drugs can block the reuptake (recycling, in lay terms) of serotonin [p129], thereby prolonging its effect at a synaptic level. They are prescribed in large quantities all over the world, even though nobody is able to explain why serotonin levels increase straight after treatment begins and yet the drug only starts having an effect several weeks later. It is to be hoped that, in the future [p230], pharmaceutical research will be able to do better than this.

9.2.3 Obsessive–Compulsive Disorder

Actions, thoughts and sometimes even words that get repeated non-stop. Feeling compelled to check five times that the front door is locked. Feeling the need to wash your hands even though you've already done it ten times in the past half-hour. Always thinking about the same thing, even if you don't want to. These are the most famous symptoms of obsessive–compulsive disorder, which, according to recent studies, affects 2.3% of the world population at least once in a lifetime, without significant variations in geography or gender.

Behavioural therapies and current medicines on the market can sometimes be effective. If there ever was any doubt about the close connection between biology and behavioural conditions, just remember the story of the young Canadian in the early 90s who shot himself in the head, survived, and woke up without any more compulsions [p149].

Evidently this condition doesn't prevent those afflicted with it from changing the world: they include theologian Martin Luther, mathematician Kurt Gödel, and inventor Nikola Tesla, who reshaped modern life as we know it.

9.2.4 Schizophrenia

Schizophrenia is a disorder in which thought, language and the perception of oneself and of reality get mixed up. Even though its causes are unknown, its effect is a general chemical imbalance of dopamine, serotonin and glutamate that interferes with the entire sensory system until it overwhelms it. Hallucinations can be visual, olfactory (related to smells), auditory, taste-related and tactile. Delusions, such as believing that your thoughts come from outside you, can dramatically interfere with everyday life.

Almost nobody develops schizophrenia as a child. Practically nobody when they're old. This condition typically manifests itself towards the end of adolescence, emerging fully at around the age of 25. Men are affected slightly more commonly than women. The risk factors are, as usual, genetic but also environmental: poverty, abuse and neglect are often connected with the onset of schizophrenia.

According to the WHO, about 21 million people world-wide are affected. The effects depend on the severity of the condition, and cases of recovery have been recorded.

9.2.5 Neurodegeneration

The good news is that the average life expectancy of brains keeps getting longer [p211], thanks to a healthier lifestyle and significant diagnostic and therapeutic improvements in health systems. On a global level, it has reached 71 years (68.5 for M Models° and 73.5 for F Models° [p178]), but with substantial geographical disparities: 83 years in Japan, 50 in Sierra Leone. Still, let us remind you that in 1900, the global average life expectancy of a brain was only 31 years, and 48 years in 1950.

The bad news is that, partly because we're living longer and partly due to improvements in medicine, in 2015, according to the United Kingdom Office for National Statistics, Alzheimer's and other kinds of senile dementia became the primary cause of death in England and Wales. This trend is forecast to gradually spread across the entire industrialised world.

Among the numerous forms of dementia that cause perma-nent damage to thinking and memory, compromising normal functioning, there are those connected to wear and tear of the brain machine, which, if detected, can be stemmed [p212]. However, things get more complicated in the case of neuro-degenerative conditions such as Alzheimer's (responsible for over 50% of dementia cases), Parkinson's, Huntington's Disease and amyotrophic lateral sclerosis (better known as ALS). Neurodegeneration, which can strike the complexity of the neural network at several levels, seems, so far, to be unstoppable and essentially incurable. It shows – in a negative way – just how extraordinary and complex the brain machine is.

9.3 DEBUNKING MYTHS

Among all the clichés, urban myths and films with low-level science content, the central nervous system is often interpreted wrongly. Please check this list of ten items of misinformation to see if any of them is already installed in your brain. If so, we strongly recommend it is removed, to ensure higher reliability and functioning of the product.

The brain sees the world as it is. Err... no.

Brain damage cannot be repaired; neurons cannot be born again; alcohol and drugs destroy them. The fact that neurons (unlike all other cells) are born and die with the brain user [p211] has engendered the belief that neural cells and their synapses are something static, if not predetermined. On the contrary, thanks to its innate plasticity [p69], the brain is able to reactivate or reallocate interrupted connections and at times whole areas isolated by a trauma. It has been proved that at least in the hippocampus (and perhaps even in the basal ganglia) some neurons continue to be born even during adulthood. Drugs that can take the reward system [p141] hostage don't produce 'holes' in the brain, as some suggest. As for alcohol, as you already know, it cheerfully interferes with neurotransmission and does not 'kill neurons', as others claim: they all get over the hangover.

Those who use their left hemisphere more are inclined towards logic, and those who use their right, creativity. Since the 1960s, we've known that the two hemispheres perform the odd different function, with preference for language on the left and spatial information on the right. However, the two hemispheres are intimately connected by the motorway of the corpus callosum and the brain operates as one single thing, not two. The notion that each brain is more dependent on the right or left hemisphere – to justify personal inclinations towards logical order or creative disorder – is alive and kicking, but has been totally discredited by science. A 2012 study proved that creative thought involves the entire encephalon electrochemically.

Mozart's sonata for two pianos K. 448 makes you more intelligent. This is a piece of news that ended up splashed on all the front pages in the world in the 1990s: it had been proved that after playing the sonata to a group of children, the results of their intelligence tests, spatial especially, would improve. No study has ever been able to replicate this experiment. And yet, in 1998, the US state of Georgia distributed a classical music CD to all children in care in order to increase their intelligence, and the 'Mozart effect' keeps being quoted as though it were real.

It's all downhill after 20. No, that's not true. Some functions peak at the age of 20, others at 30, others – language, for instance – at 40. Maturation and mental decline [p212] are much more complex than we think because different factors influence different cognitive faculties.

Crosswords and sudoku keep the brain in shape. No, that's not enough. Crosswords and software programmes, advertised as a defence against brain ageing, simply increase the ability to solve crossword puzzles and quizzes. They are a memory test, but do not increase intelligence. They're certainly useful as long as they're not viewed as a cure-all. Constantly learning new things outside one's comfort zone [p162] is probably more effective. Putting it simply, you need to put in some elbow grease.

Mirror neurons have shaped human civilisation. In the 1990s it was discovered at the University of Parma that motor neurons in monkeys are activated not only to trigger a certain movement but also when they see anybody else making that same movement. Hence the name mirror neurons. Indian neuroscientist Vilayanur S. Ramachandran's theory on the subject gave rise to some exaggerated ideas. The theory – now widely contested – attributed to them the function of 'empathy neurons' [p132], responsible for human civilisation and, in the case of malfunction, for autism. Nowadays, we sometimes read quips like 'It's mirror neurons that make us cry at the cinema' or 'It's beneficial to go and visit friends in hospital so as to activate mirror neurons'. Without in any way diminishing the importance of the discovery, recent studies highlight the fact that mirror neurons are part of an intricate network of neural activity, including empathy, which naturally avails itself of the imitation function. But they are not its 'on' switch.

There are people who can read minds or use extrasensory perception. This myth, which took hold in the 1930s, consists in believing that the brain is capable of perceptions that don't come from the senses but are produced by the actual mind: from infallible intuition to clairvoyance, to telepathy and even telekinesis (the ability to move objects with your thoughts). Science, the human activity that deals with understanding nature through verifiable and repeatable observations, rules out the possibility that there could be any truth in it. However, there are those who believe the contrary, even in the higher echelons. The story of that group of 'psychic spies' organised by the CIA during the Cold War in order to test military solutions based on thought is very instructive. And the film made about it, *The Men Who Stare at Goats*, is very entertaining.

We use only 10% of our brain. Hollywood encouraged this old urban myth with the film *Lucy*, in which Scarlett Johansson accidentally ingests a huge dose of nootropics [p230] and her intelligence increases every hour until she gets endowed with telepathy, telekinesis and mental teleportation. Fantasy. The truth is that the brain is already engaged 100% in making the entire human being function: breathing, heartbeat, blood pressure, digestion, movement, balance, thought, future planning and so on. Watching television while eating crisps – which could seem like the height of 'doing nothing' – actually consists of neural work to all intents and purposes. Even during sleep, the brain is fully active. This 10% business is the joke of the century.

The quantum mind. Some, like physicist Roger Penfield, have put forward the theory that quantum mechanics plays a determining role in many cognitive processes, starting with consciousness. In other words, the possibility that brain functions are halfway between the standard model of physics that *rigidly* regulates the world around us, and quantum mechanics, which *probabilistically* regulates the subatomic world. It's a theory that's very hard to demonstrate: 'If you think you understand quantum mechanics,' US physicist Richard Feynman famously quipped, 'then you don't understand quantum mechanics.' In the fullness of time, the theory may either be proved or refuted. In the meantime it's fuelling a host of pseudo-scientific theories about the inherent individual ability to improve one's life by tapping into the 'quantum mind' and even cure illnesses with a 'quantum cure'. Hmm.

10.0 END OF LIFE (EOL)

ALL PRODUCTS THAT COME WITH A user's manual have a limited, albeit undefined, lifespan. Some people call it **built-in obsolescence**, implying deliberate responsibility on the part of the manufacturer. In the case of the brain, it is programmed by nature.

Skin cells live approximately a month. Red blood cells are renewed every three months. The cells of the liver every eighteen months. Neural cells, on the other hand, last the entire lifespan, so that elderly people can retain memories of their childhood and thus remain the same person every passing day. This peculiarity makes neurons the real heroes of existence.

It's only in the past few years that we've debunked the myth that between fifty and eighty thousand neurons die in the human brain every day and are never replaced. We now know that as we grow old the encephalon loses a certain number of nerve cells, but we also know that it's able to produce new ones (in the hippocampus in particular) [p45] throughout life. This doesn't mean, however, that the road through old age is toll-free.

Old age affects the brain at a molecular, cellular, vascular and structural level. Even though the mechanisms are still not fully understood, genetics, daily experiences and also variations in the levels of neurotransmitters and hormones play a part in gradually damaging memory, motor ability and executive functions. With people living longer, we have a greater opportunity to witness neurodegeneration [p205].

The loss of synapses during adulthood contributes to a gradual decrease in the density of the cortex. Between the ages of 60 and 70, grey matter slowly starts to get thinner, especially in the frontal lobes and in the hippocampi. This is equally true for so-called white matter, because the myelin sheath coating the axons deteriorates. The elderly brain produces fewer neurotransmitters and possesses fewer receptors to receive them. Lower dopamine, serotonin and acetylcholine [p20] levels contribute to the loss of memory and sometimes to depression. If we add that wear and tear to the blood vessels, together with high blood pressure, increases the risk of a stroke, we can see that in comparison with obsolescence of the central nervous system, wrinkles are nothing.

In Buddhism, old age is one of the four sufferings of existence. The others are, chronologically, birth, illness and death, which is tantamount to saying that problems tend to concentrate in the latter part of life. Being consciously prepared for it (and prepared doesn't mean anxious) can prove to be truly beneficial.

With an adequate phase of approach and a few common-sense strategies [p224], we can keep the general effects of brain deterioration at bay. We recommend you prepare yourself well in advance to tackle it.

10.1 APPROACH PHASE

Growing old gracefully is often good fortune, a chromosomal gift from our forebears. No wonder scientists are searching for the lucky key in the genome of populations where ultracentenarians are more common, such as in Ogliastra in Sardinia, in the north of the island of Okinawa in Japan and in Ikaria in Greece.

They hope, for instance, to find a way of slowing down the deterioration of telomeres, the repeating 'words' at the end of each of the 23 pairs of chromosomes: hundreds of 'TTAGGG' written in the alphabet of the four nitrogenous bases that make up DNA. Telomeres are there to protect the genetic heritage during the replication of DNA, but, because of the mechanism used, they lose segments over time. The fewer they lose, the slower the ageing. Succeeding in protecting them would be a triumph for medicine and a disaster for national pension schemes.

Growing old gracefully is not only an inheritance but an art. Having the right genes does not guarantee a long and healthy existence if one's lifestyle is terrible. Centenarians, and those who keep themselves in good shape in general, have a personal (possibly unconscious) formula for physical and mental health. Halfway between science and mere common sense, the wisdom garnered by watching super-agers, the super-heroes of ageing, could be summarised in five points:

Movement. As early as start-up [p79] both brain and body – the *mens* and *corpore* in the Latin – need physical activity to develop and function correctly [p91]. Walking a lot, cycling, doing frequent and even heavy gardening work seem to be closely related to a long life. We're not talking about the gym here (which is, of course, an excellent alternative to immobility), but about a daily and established habit of moving.

The elderly residents of Ogliastra, Okinawa and Ikaria – called 'blue zones' by Dan Buettner, a researcher at the National Geographic Society who studies the geography of longevity – have walked kilometres every day of their lives: a habit more than a duty. Seventh Day Adventists, who underwent a similar study in California because of their longevity, traditionally spend every Saturday of their lives walking in the fields and the woods: a pleasure more than a duty.

In short, always moving is not just a recommendation for the old-age-approach phase but for life as a whole. The sooner one starts, the better.

Nutrition. Energy comes with eating, but so do illnesses. According to the World Health Organization, in 2016 there were 1.9 billion overweight people on the planet, one third of them obese. Putting it bluntly, there isn't a single overweight super-ager. The longest-living populations are characterised by a chiefly vegetable diet, low in animal fats, with micronutrients, omega 3 fatty acids and antioxidants (such as Japanese green tea or Sardinian Cannonau, which apparently has more polyphenols than any other wine).*

Studies on monkeys have shown a strong connection between calorie restriction in diet – which is very different from malnutrition – and delayed ageing. Therefore, in addition to following basic suggestions about healthy eating and hydration [p84], it would be desirable, during the phase of approach to old age, to pay a little more attention to the substances inserted into the digestive system, and above all to moderate their quantity. They say that in Okinawa they have the custom of 'filling the stomach to 80%'. It may be difficult to calculate percentages with the belly, but this is a perfect example of an attitude that aims for healthy calorie moderation.

Meaning. The work–pension social model, seen as an abrupt transition from activity to inactivity, is a disaster. It's no

---

* The consumption of red wine is linked to longevity but only if drunk in moderation.

wonder many people look for other jobs or interesting activities to give their daily lives a meaning. Those who don't tend to deteriorate much faster physically and cognitively. In the blue zones, there is a family system based on an almost sacred respect for old age, considered to be the age of wisdom, in which the old person plays a part in caring for descendants and has a meaning in life.

In order to age slowly, your life needs to have a goal, as suggested by the Japanese notion of *ikigai*. *Iki* literally means 'life, existence' and *gai* 'result, effect, fruit'. It can be translated as 'reason to live'. In Japanese culture, everybody is expected to seek his or her own *ikigai*, combining individual preferences and choices, and give it free rein in the post-working age. According to Buettner, however, they are even more specific in Okinawa. On that island in the south of the Japanese archipelago, *ikigai* can be translated as 'the reason for getting up in the morning'.

We cannot give you indications about your own *ikigai*, because it depends on personal choices. Suffice to say that, because of sophisticated neurotransmission mechanisms, depression engenders more depression [p202] whereas motivation engenders more motivation [p154]. A meaning to life that goes beyond your job (such as after you retire) and the phases of your existence (such as when children fly the nest) is essential to the approach phase. It's a matter of finding the answer, the one that makes the most sense to you, to the question: what do I wake up for in the morning?

Socialising. The positive correlation between ageing slowly and relationships with other people is overwhelming. Super-agers have in common not only physical exercise, nutrition and the motivation to get up in the morning, but also a close rapport with relatives, friends, and other humans, too. A large number of psychological studies show that activities such as volunteering, belonging to a religious, cultural or artistic group and regular visits to theatres or recreation clubs considerably reduce the potential negative symptoms of an ageing brain.

Although on the longevity islands, an old person gets special treatment and has a role, in the rest of the world – especially the urbanised world – those who become old often turn into a problem to be solved. If we think that the average lifespan is destined to increase [p205], doing all we can to age slowly seems almost imperative. Therefore, if you prefer solitude, may we suggest you rethink. It's bad for your health, both general and cerebral.

Knowledge. The fifth point is not inferred from observing populations with notable longevity. Nonetheless, many studies confirm that there is an inverse connection between the rate of education and neural deterioration. It therefore deserves separate consideration.

10.2 LIFELONG LEARNING

A noteworthy scientific study begun in 1986 opened a window on Alzheimer's, that most dreaded age-related illness. A group of researchers from the University of Minnesota compared texts written decades earlier by 700 novices of the Sisters of Notre Dame, a religious order, with their medical records when they were old. They found a close connection between higher cultural attainment and a reduced tendency to develop this illness.

If learning and knowledge represent a protective shield against senile dementia, governments had better seriously support and promote them. Now that the army of baby boomers (born between 1946 and 1964) has begun to go beyond the 70-year threshold, the WHO expects the incidence of Alzheimer's to multiply threefold by 2050. The social cost of this anticipated epidemic will be extremely high. A timely intervention, even among younger generations, could prove to be the shrewdest investment.*

The 1996 Delors Report (named after the famous President of the European Commission) sent up the first rocket. This inspired document suggested disseminating the idea of a lifelong learning principle across the educational system. The idea was to allow people to carry on developing skills, knowledge and personal attributes facilitated by an education system capable of lasting an entire lifespan and based on 'four pillars': 'learning to know', 'learning to do', 'learning to be', 'learning to live together'. An idea backed up by common sense, by economic interests and, as we know nowadays, also by science. And yet it remained a utopia.

We recommend that you turn it into personal reality.

Lifelong learning does not mean dealing with school bells, exams and reports all your life. It simply means following a path of neural, synaptic and axonic [p11] growth that's entirely voluntary and entirely free. It means tapping into the human tendency towards curiosity in order to guide, at will, the brain's vigorous neuroplasticity [p69] so as to shape a more aware citizen, a more flexible worker and a brain more resistant to ageing.

To quote the title of an old Italian TV programme dedicated to eradicating illiteracy: *It's never too late*. State schools, along with state television, have managed to close that embarrassing

---

* The WHO estimated (in 2014) a global cost of 607 billion dollars a year, including the lost working hours of those who have to care for Alzheimer's sufferers.

social divide. Never has it been as easy as it is now, thanks to technology, to achieve the next step: lifelong learning.

Whether you're an energetic teenager or a white-haired baby boomer, it's never too late to start having fun. Because lifelong learning implies fun. The freedom to decide what to learn and know, what to learn to do, to be, and what to learn from others. It's the reward system [p141] that drinks a dopamine toast to the pleasure of constantly adding new modules to its neural warehouse of knowledge.

You could decide to add the 'watercolour' module this year and start on the 'Spanish' module next year. You could program software, learn billiards, Chinese cooking, the oboe, Sanskrit. Anything you like. On the Internet, there are videos on how to play the ukulele, there are books and encyclopaedias containing the entirety of human knowledge, and university courses available totally free of charge. All that and much more is accessible on the electronic device you carry in your bag or in your pocket. History's most famous lifelong learners – from Socrates to Leonardo da Vinci – would turn green with envy.

Like any other tool, this ocean of digital information called the Internet can be used in a – let's say – more or less intelligent way [p72]. Exploited as a tool for constant self-education, it represents a milestone in the history of planet Earth. Until the early 1900s, centuries after the invention of the printing press, books were still the prerogative of the aristocracy, the wealthy and religious institutions. In the 21st century, there's not a single barrier left in the way of acquiring information.

It's the perfect time to embark upon the path of lifelong learning. If possible, until the very end.

10.3 AFTER ALL

Your brain is 100% biodegradable. We regret to inform you that the entire contents are deleted just a few seconds after the machine has reached the end of its run. Since it's the largest store of information about your experience on this planet, losing it for ever could be unfortunate.

While waiting for technological advances to allow us to make brain back-up copies [p234], we recommend that you record those details you consider to be most significant, so you can pass them on to whomsoever you wish, usually your grandchildren.

Every brain has its unique, unrepeatable, very often secret story. Who wouldn't like to read the intimate thoughts and personal stories of their father, grandmother, or even a distant forebear?

This story can be recorded in various formats – text, audio or video (in ascending order of emotional impact on the recipients). The video solution can actually produce undesirable effects. For this reason, we strongly recommend text format, possibly nicely furnished with old photographs.

If your brain has spent the last few years or decades incessantly 'posting' thoughts, words and images on Facebook, Twitter and Instagram, then let's just say: Thank you, that's all right. Perhaps that'll do.

11.0 EXTENSIONS

REPLACING THE CENTRAL MICROPROCESSOR WITH A more modern one, mass memory with a more capacious one, updating the operating system. With these few actions, an old computer can get a new lease of life and process calculations more smoothly than ever. There are, however, no commercially available expansions or upgrades for the brain. Still, don't think that biology is in any way inferior to electronics.

The central nervous system knows how to self-assemble, connect, rebuild itself, and, in some cases, self-repair and even evolve. In short, it has its very own way of adding neural structures to its cognitive heritage and in this way updating the management of the entire system and the installed applications, as well as contributing to general smooth running and slower obsolescence. These kinds of expansions are not commercially available for computers.

As a general rule, brushing your teeth with your non-dominant hand – or anything else that allows you to leave your comfort zone [p162] – creates new synaptic connections and perfects motor and cognitive control.

But much more can be done. The bounds of memory can be broadened, entire neural prairies can be added by learning a new language, great fulfilment can be arrived at through meditation and a fundamental sense of gratitude for life itself. Or, if we're willing to cheat a little, we can always resort to the doping of the mind. Medicines and food supplements are currently used to increase cognitive and attention abilities, often altering the status quo in competition between students or work colleagues. It's understandable that this quest should be embarked upon by a living species which, albeit primitive, is intelligent enough to value intelligence.

11.1 EXPANDING MEMORY

It is the gradual appearance of cars, scooters and lifts that has made modern gyms a necessity. While technology has added speed and comfort to the life of the human species, it has also removed a large proportion of the muscular activities that had been accompanying it evolutionarily for hundreds of thousands of years [p91].

And what about memory? It has been an essential resource in the evolution of *sapiens*. Not only to carry the necessary practical baggage (which plants can be eaten and which avoided) and the social baggage (the so-called oral tradition), but also to accumulate the culture and knowledge necessary for creating new ideas from old ones, and to then decant them into other brains. History tells us for example the legend of King Cyrus of Persia, who knew all his soldiers by name. Or of the encyclopaedic knowledge of Renaissance philosopher Pico della Mirandola. It was in those past ages that memory had huge importance and prestige.

With the arrival of books, things began to change: there was no need to memorise perfectly a list of plants, formulas or elements, as long as you kept the manual nearby. As early as the mid-eighteenth century, philosopher David Hume was lamenting the passing of the times of Cicero and the heroes of Antiquity, 'when the faculty of memory was of paramount importance, and was valued much more than now'.

What can we say about the situation three centuries later? The first mobile phones made memorising the phone numbers of friends and relatives unnecessary. The new smartphones store appointments, addresses and messages, and clear any doubts about the first name of a singer or politician on demand. With Bluetooth technology automatically connecting them to your car, you no longer need to remember where you've parked it. Then there are virtual home assistants on the market that, upon vocal request, remind you of the dates in your calendar or what that song is called. Some predict that by 2020 there will be an average of 6.5 connected digital devices for every human being (it's what is collectively called the **Internet of Things**), including thermostats, videocameras, watches and eye-glasses. In this triumph of digital memory, what will become of biological memory by the middle of the century? Or by the beginning of the next one?

This risk is confirmed by the widespread tendency to shrug one's shoulders: 'I have a memory like a sieve, I can't help it.' Or, even worse, by the belief that memorising something in some way takes space away from other memories, when it's the contrary that is actually true [p69]. The fact is that memory is at the core of the most extraordinary function your brain possesses, learning [p162], and actively using it contributes to psychological and physical wellbeing and slowing down ageing [p216].

Doing this well is not a question of what type and quantity of information to memorise. It's a question of how you go about the process.

Anyone who went to school in England, China, India, Japan, Brazil or Turkey regularly had to commit a poem to memory. A commendable effort to shape the mnemonic skills of a young and plastic human brain, but also mechanical and ineffective. This task works against what (little) we now know about the mechanisms of memory. A text learnt through repetition is boring, does not help us to understand, does not produce associations with other knowledge and, in time, fades.

Anyone who wants to learn to recite *The Divine Comedy* or to play *The Well-Tempered Clavier* by heart cannot resort to mechanical repetition. They must use a method that appeals to passionate attention, a deep understanding of the meaning (whether literary or harmonic), and a sequence of mental associations that favour the neuronal retrieval of the word or note sequence. Interestingly, in English we refer to memorising 'by heart'. However scientifically inappropriate, the idiom emphasises that without passion – something more than simple attention – not much can be memorised.

There are many books and courses that teach more or less effective mnemonic techniques. And yet, if brains would only drop the widespread prejudices about the limitations of their memory, that would already be a big step forward. Memory isn't a 'muscle' to flex with crosswords, as some say [p206], but an intricate electrochemical exercise to strengthen synapses and axons. Everybody has plasticity. Everybody can have the memory they wish to have [p64].

It takes ten minutes to expand the memory of a computer. In order to expand human memory, one must cherish the old memory and add information to it for ten years (usually the time necessary to become an expert at something). Even just to compensate for the de-memorisation of phone numbers and addresses, it would be good if you transitioned to the gradual memorisation of something else, whatever takes your fancy. The lyrics to all the Beatles' songs? The list of atoms in the periodic table? The 194 capital cities of the world? The sky's the limit.

American journalist Joshua Foer found exactly that when,

after interviewing a few memory world champions for an article in the *New York Times*, he decided to test their suggestions for memorising on himself. The story he tells about it in his book *Moonwalking with Einstein* has a most definite happy ending.

To commit the fateful poem to memory, it's much more effective to use associations – visual or emotional, for example – than mechanical repetition. German memory athlete Gunther Karsten suggests anchoring key words in mental images that are as absurd or ridiculous as possible, in order to increase effectiveness. Corinna Draschl, another memory champion, uses associations with specific emotional states (perhaps because she is equipped with an F Model® brain) [p178].

Foer says that he simply used the method, well known in Ancient Greece, called the Memory Palace. This consists, in short, of choosing a large and familiar place, for example your grandparents' house in the country, and going through it with your mind from one room to the next, placing on the sofa, the table or in the corner of a kitchen a series of mental objects associated with the sequence one means to memorise. And it's thanks to this ancient technique that, in 2006, Foer won the USA Memory Championship after training for just one year, memorising in one minute forty seconds the exact sequence of 52 playing cards.

So don't let anybody say that memory cannot be expanded.

11.2 STRATEGIES FOR THE BRAIN

The first strategy for the brain is to follow the basic recommendations for its exercise and maintenance. You can be as intelligent as you like, but without an adequate supply of food and hydration, sleep and movement, your brain will jam just like an internal combustion engine without lubricant [p93].

With this proviso, here are some strategies that may contribute to expanding the brain's ability and potential, as well as delaying its ageing [p212].

Reading. With the democratic and egalitarian opportunities it offers for indefinitely enlarging the pool of knowledge, the Internet represents an historical turning point [pp162, 216]. And yet its thousands of exchange and sharing ramifications also represent a powerful tool of mass distraction, such as the double-edged sword of multitasking [p162].

In a 2008 article published in *The Atlantic* titled 'Is Google making us stupid?', writer Nicholas Carr lifted the curtain on a possible and worrying consequence: the theory that the Web's very hypertextual nature could have reverse neuro-plastic effects, narrowing the brain's ability to contemplate and concentrate. As proved by the forming of habits and addictions, plasticity doesn't always have a positive conse-quence. According to Carr, who laid it on thick in his book *The Shallows: What the Internet is Doing to Our Brains*, the immediate result of this cerebral process is that people find the long and meditative reading of bygone days less appealing.

What does your brain think of this? In the past few years have you read fewer books from beginning to end, by any chance? Are you more attracted to the speed of hypertext, of Wiki sites and of social networks? Calculate this for yourself.

The book, whether in paper or virtual form, contributes to shaping the architecture of reflection and 'slow' thought. The Wikipedia-style hypertext, conversely, contributes to the building of mental associations between one fast piece of information and another. The book carries with it the (possible) authority of the author and publisher. Whereas the hyperlink assumes you can make a distinction between the varying (and at times unknown) quality of the sources: being unable to do so means risking being overwhelmed by the wave of 'fake news' that has proved its ability to toxify international democracy and diplomacy.

The second strategy for expanding the boundaries of the brain can therefore be articulated in three points: reading, reading, reading. Reading is good for the synapses, from childhood to the third age.

Meditation. Another excellent strategy for expanding the brain's capacity for concentration and contemplation is 'mindfulness' meditation. The term, coined recently but derived from an ancient Buddhist tradition, indicates the process of channelling all your attention to the present moment. Feeling every nook and cranny in your body, the weight of your feet on the floor, your breathing, the thoughts that run through your head, and, indeed, the passing moment.

For about ten years now, mindfulness meditation has been accepted as a therapy to combat anxiety and depression; it is also being explored as a means of relieving the symptoms of other conditions. It is being tested in schools, gyms and barracks as a strategy to improve educational, sports and psychophysical results. Between the beginning 2016 and the middle of 2017, 7,820 scientific papers were published that mention or hint at the effects of mindfulness meditation on the brain. Some English language sites that deal with it receive millions of visitors. Since there are now courses, schools and self-help methods that recommend it, it's up to you to select the most sensible advice (and it's possible to pursue it without spending a penny).

Magnetic resonance scans reveal that, after just eight weeks of meditation practice, the amygdalae (the centre of fear) [p44] get smaller while the cerebral cortex [p51] grows thicker. No wonder it's been proved that meditation lowers stress and increases attention.

Ideally this should be a daily exercise. Practising meditation so that it has an effect means carving a small space for it every day, perhaps just a few minutes to begin with. Turning all your attention to the present moment means just that. Multitasking and mindfulness don't go together.

Music. There's a fascinating connection between the brain and music. Even though the evolutionary reasons for this are unknown (it doesn't help us to procreate or survive), a beautiful song or a string quartet increases dopamine and decreases cortisol. Listening to a Mozart sonata does not make you more intelligent [p106], but a plethora of psychological studies and tests with functional magnetic resonance still confirm that music is a very powerful strategy for expanding the brain.

Thanks to the aforementioned neurochemical effects, music can alter one's mood, energise, and in some cases help with concentration. Workers who are able to choose the music they listen to are more productive. Listening to music, a favourite human activity, triggers a vast network of brain structures that enable you to distinguish the pitch of the sounds and perceive the rhythmic and harmonic structure of the composition. Almost nobody listens to a record in religious silence, as people once did, but now that music is digitised, streamed and always available, listening rates are higher than they've ever been in the history of humanity.

It's by *making* music that we achieve the greatest gains in brain expansion. Musical education in childhood (the correct critical period) improves verbal and reasoning abilities because of a plastic modelling of the auditory, motor and sensory-motor system. Scientific research has found differences at the cognitive, structural and functional levels between the brains of musicians and non-musicians, although it's difficult to prove that these depend exclusively on the ability to play an instrument.

Ideally one should start very young. However, although more difficult, it is equally beneficial (and enjoyable) starting in adulthood. If you are one of those people who learned an instrument as a child and then, for whatever reason, put it away, we recommend that you do not throw away the brain heritage of a strategy as powerful as music.

Languages. Speaking more than one language represents a distinct competitive advantage in this globalised age. They say that there are more bilingual than monolingual individuals in the world. Certainly, over half of all European citizens speak at least two languages. India has established its 23 official languages by law. In Mumbai and Kolkata, it's not unusual for someone to speak Punjabi and Hindi with their father's family, Bengali with their mother's, and English with their children. Science tells us that this is all very good for the brain, to the point where it increases its plasticity potential [p69].

The critical period [p79] for learning languages is early childhood. Even early on, children easily learn three languages at the same time, for example their father's, their mother's and the one spoken at kindergarten. If one of these then ceases to be used, when they grow up they will still be able to distinguish its phonemes better than other people.

Scientific studies have registered superior cognitive abilities in bilingual brains that go beyond the sphere of languages – and naturally, it's easier for them to learn a third or fourth language. If this weren't the case it would be impossible to become a polyglot like Emanuele Marini, an office worker in the Milan area who speaks sixteen fluently.

Those lucky enough to be born bilingual should hold on

to that. Those lucky enough to be very young should be given the chance to immerse themselves in a non-native tongue. Those lucky enough to be older and wiser should seriously consider the strategy of expanding the boundaries of their brain by learning another language.

Gratitude. Luck, indeed. Life is a stroke of luck and everything seems to suggest that just thinking this is a successful brain strategy. This discovery has been confirmed by several tests. In one of these, a group of young adults was asked to keep a daily diary of things for which they were grateful, a kind of reflection on the positive sides of their lives. Others were asked to keep a diary of things that went wrong, and a control group to keep a normal diary. The result, after as little as two weeks, was that higher levels of attention and more enthusiasm and energy were recorded in the first group. Gratitude is good for your health and it's a well-known fact that, when regularly applied, it is also highly beneficial to human relationships.

'*Gracias a la vida*,' the song of the same title goes, immortalised by the Argentinian singer Mercedes Sosa, '*que me ha dado tanto*'. Although 'positive' psychology can seem strange or implausible to some, there is good evidence that it works [p122]. The gratitude experiment has been repeated in multiple iterations, for example asking the subjects of the experiment to keep a diary just once a week. This produced the same results, which were corroborated by functional magnetic resonance [p234]. The beneficial effects of appreciating life – which is clearly the exact opposite of brooding or complaining – appear to be long-lasting and self-reinforcing: just as brooding too much produces self-perpetuating negativity, giving thanks is a virtuous circle that lightens one's outlook. It's not a matter of shutting your eyes to problems, but observing them from the correct perspective.

It's not hard to try. All you need is to feel grateful for the beauty of the world and the good fortune of having a brain capable of seeing it.

11.3 MOLECULES FOR THE BRAIN

NZT-48 is an incredible drug used for cerebral enhancement. No wonder it has multiplied its inventors' initial investment tenfold. To be precise, it has made 236 million dollars for the producers of *Limitless*, the film with Bradley Cooper and Robert De Niro (which cost 27 million to make), in which a loose-living writer becomes a genius simply by swallowing a nootropic pill. One can deduce from its box-office success that the dream of becoming much more than intelligent tickles the secret fancy of audiences all over the world.

Nootropics are drugs that enhance the cognitive abilities of the human brain. None of them has even remotely the power of the Hollywood drug, and yet they are already a commercial and cultural phenomenon. Cultural, because they are mainly taken by students in the most famous American universities and by employees at the most famous companies in Silicon Valley. As for their commercial success – that's evident from the widespread availability of nootropics sold on the Net.

Naturally, there are molecules and then there are molecules. The most potent drugs are those that require a medical prescription and which, according to the news, are procured more or less illegally in order to obtain a competitive and not very ethical advantage within a company or a university. Methylphenidate, sold under the name of Ritalin, is used to manage attention disorders, such as ADHD [p157], on a large scale in the US (it's also sold in the UK, but not in Finland).

In off-label use – outside the pharmaceutical company instructions – Ritalin improves attention and concentration, boosts energy and enhances brain performance in difficult or repetitive operations. Adderall, prescribed for both ADHD and narcolepsy, is apparently used in the workplace, professional sports changing rooms and bedrooms. It adds strength, drive and a touch of euphoria, which sounds rather good.

And yet these drugs can cause harm. Besides the possible side effects, if used over a long period and in high doses these molecules easily trigger strong physical and psychological dependency – Adderall in particular, which is not sold in Europe and is classified as part of the amphetamine family to which it belongs almost everywhere in the world.

There are other nootropic drugs on the market. There's modafinil (known as Provigil), which is sometimes used off-label to ease depression and cocaine dependency. There are two classes of molecules prescribed to people with Alzheimer's which are thought to develop the cognitive faculties of a healthy brain. Not to mention the racetam category, such as piracetam (sold in Europe under the names Lucetam and Nootropil, but banned in the US), which is often taken for neuronal enhancement, though its pharmaceutical licence is, in fact, for the treatment of muscle twitches.

There is another category of nootropic products flooding the market, commonly referred to as 'stack'. They are classified as food supplements and do not require a medical prescription. Indeed, in a pill they stack a variety of molecules – either natural or derived from natural sources – in order to create an ideal blend to improve cognitive function. In June 2018, searching for 'nootropics' on the UK Amazon website, fifteen or so products came up, and there were almost 1,500 on the American site. Reviews are helpful, of course, but choosing between Mind Matrix, Neurofit, OptiMind and other bottles of pills feels like a monumental task. Besides, stack aficionados usually mix them until they find the ideal, personalised stack. Naturally, the contribution of these pills to cerebral function is not as dramatic as that of the pills on the silver screen. Their effect is only just perceptible, even though users claim that results kick in more strongly after some time. In other words, nothing like the instant intelligence of NZT-48.

This doesn't mean that the pharmaceutical industry, capable of investing hundreds of millions of dollars in researching a single molecule, isn't already hoping to develop nootropic products that come closer to the collective dream of *Limitless* but without being addictive or involving unpleasant side effects. It would be the perfect smart drug, able to rake in astronomical figures. Just remember that the two mildest nootropic substances in the world – caffeine and nicotine – turn over hundreds of billions of dollars a year.

It's just a matter of waiting for the future to become the present.

12.0 FUTURE VERSIONS

LIKE ALL THE MANUALS IN THE world, this one makes no attempt to reveal secrets regarding versions that will appear in the future.

Well, actually, to be honest, it would like to. The only problem is that nothing's harder to pin down than the future: a manual on intelligence isn't so foolish as to try.

Nonetheless, in defiance of all the manuals in the world, it will take the liberty of offering a few reflections on how intelligence could evolve in future system versions. Reflection does not mean prediction, after all.

The brain has taken hundreds of millions of years to evolve from the ancestors of reptiles to the descendants of Leonardo da Vinci: at this rate, you wouldn't expect much over the next couple of hundred years. However, thanks to the advances in neuroscience – not only genetics but also micro- and nano-electronics – the advent of machines capable of increasing cerebral potential, genetic technologies to prevent neurodegeneration, and machines to replicate and exceed the levels of average human intelligence, seems practically inevitable.

In a public appeal, four influential scientists, including Stephen Hawking, wrote: 'There are no fundamental limits to what can be achieved: there is no physical law precluding particles from being organised in ways that perform even more advanced computations than the arrangements of particles in human brains'. Erudite words which basically aim to raise the alarm: there is a risk that someday we might create an intelligence so superior to human capability as to make humanity itself redundant.

But is this a risk that belongs to the immediate future or to a vague, distant future?

12.1 NEUROTECHNOLOGIES

Distant Past. Let's forget about the most rudimentary methods used in the old days for studying the brain: from applying electrodes, as pioneered by Luigi Galvani in the eighteenth century, to the drill and hacksaw. This story can begin in 1924, when the first human brain underwent an **electroencephalogram**: at last a non-invasive technology to give us an idea of what's happening on the inside. This is how – through a network of electrodes applied to the scalp – they discovered neural oscillations, better known as brainwaves [p12]. Almost a century later, a vastly more refined version of that primitive recorder of neuroelectric impulses is still used in medicine as well as in research.

Recent Past. The leap in quality truly begins in the 1970s with a real crop of neurotechnological inventions and discoveries. **Computed tomography** (CT) appears, which uses x-rays to produce cross-sectional images of the body that are then processed three-dimensionally with an algorithm which, in its earliest versions, could take three hours to compute.

Then we see the appearance of **magnetic resonance imaging** (MRI), which uses magnetic fields and radio waves to represent internal anatomical images. **Positron emission tomography** (PET), already thought of two decades earlier, becomes reality: through a contrast medium, we can observe physiological and not just cerebral functions. With **magneto-encephalography** (MEG) we have been able to start drawing a map of the brain, thanks to magnets so sensitive that they can detect faint neuronal activity. In other words, a true technological arsenal – albeit initially very primitive – to fathom the invisible depths of the central nervous system without using a hacksaw, a drill or even electrodes.

The Present. The evolution of hardware devised decades ago has triggered in the new century a veritable neuroscience revolution that is still yielding abundant results. All these technologies are being constantly fine-tuned as well as further developed, thanks to microprocessors' burgeoning capacity for calculation. The CT scan is no longer axial, the PET has also become the SPECT (single-photon emission computed tomography) and MEG magnets have reached unimaginable levels of sensitivity.

But magnetic resonance has become the real star on the global neuroscience stage – ever since the 'functional' label was added to it. **Functional magnetic resonance imaging** (fMRI) is able to show the most active areas of the brain in real time and in three-dimensions: since these active areas require the most oxygen, the trick is to trace the movement of the blood that carries it. Most of the recent discoveries listed in this manual derive from this technology, though not always from that alone.

Every neuroimaging technology has its merits and flaws, but the flaws can often be compensated for by combining technologies. In recording the course of brain activity through time, MEG offers a precision of ten milliseconds, while fMRI has a temporal resolution of a few hundred milliseconds: this is why, depending on circumstances, they are used together or in combination with other technologies. Although they still look primitive, their fields of application are already the stuff of science fiction. Just to give an example, fMRI has already been used in a few police investigations in order to determine the degree of awareness (and therefore guilt) of violent criminals.

The Future. Starting from zero less than a century ago, neuro-technology has taken giant steps. What everyone expects is that, in the future, it may actually enhance the brain.

If this makes you think of totally futuristic technologies, like microchips that interface with the brain, or cranial stimulation that increases cognitive abilities, then we are happy to inform you that these already exist. **Neuronal implants** have been designed that can treat people with severe epilepsy by inhibiting activity in certain areas of the brain, or that can allow those rendered paraplegic to move artificial limbs with their thoughts. **Transcranial magnetic stimulation** (TMS), capable of modulating neuron excitability without invasive methods, is already being used for research, and in serious cases of depression and neurodegeneration.*

At the moment, in order to activate a neuronal implant, a door to the brain needs to be *physically* opened. This dramatic procedure is only likely to be used on someone affected by epilepsy, amnesia or paralysis. But past experience – from MRI to mobile phones – suggests that in the space of 30 years technical and technological progress can be literally unimaginable.

We know that technology follows more or less the same evolutionary phases. In the beginning, experiments on the brain–computer interface will be rudimentary, with the odd

---

* There is also transcranial direct current stimulation (tDCS) which bathes specific areas of the brain in a low current. These kinds of devices, which are applied to the scalp, promise to improve the efficiency of the brain and are already available to buy online. Their effects and consequences are yet to be proved.

serious problem and some other contra-indications. Then, gradually, they will begin to improve until they reach the threshold of large-scale marketing. Subsequently, two or three versions later, bioelectronics will have evolved sufficiently to treat many disorders, perhaps capable of improving memory, concentration and even mood.

The road to this level of technology may seem very long. Even though, in barely a century, research has acquired a wealth of extraordinary instruments, science is still far from knowing all the structures, connections and functions of even a single brain. Just think of the human genome, which was sequenced over fifteen years ago. Most of the genes it contains have been identified, but the way in which they work together remains a mystery. Not to mention the differences between different people's genomes, also largely a mystery.

The plan now is to decipher the **connectome**, i.e. to draw a precise map of cerebral connections. It is such a strategically significant project that the US have launched the Brain Initiative and the European Union the Human Brain Project, two ten-year research programmes – multidisciplinary and lavishly financed – in order to produce a kind of brain atlas. It's an open secret, already known to everybody, that once this is completed, the brain will still be an enigma.

Forget about the distant future, the one in which our descendants will be able to download their brains and – just as science fiction used to imagine and science now fantasises – be able to live for ever inside a computer that's much more sophisticated than those we have now. Maybe at that point, we'll also be able to defrost the brains of those ex-billionaires and ex-optimists who, since the 1990s, have been getting themselves frozen in the hope that someday technology will become so advanced that they will be brought back to life, ideally more intelligent and more full of spunk than before.

Anything can happen, but these things truly are too distant to truly conceive.

Nearer at hand, maybe a matter of 30 or even 60 years, the platoon of institutions busy deciphering every detail of the human brain and its extraordinary complexity (among which we should mention DARPA, the science branch of the Pentagon) will inevitably open the way to new and powerful neurotechnologies, which may transgress ethical boundaries. From the modest starting point of the knowledge we have nowadays, this might seem like an impossible enterprise. Yet this same knowledge is sufficient to tell us that, theoretically at least, there is nothing to prevent the **artificial evolution of intelligence** from becoming real.

This truly would be a huge, historic upgrade (4.3.8) of the system version [p10].

12.2 GENETICALLY MODIFIED BRAIN (GMB)

For thousands of years, *Homo sapiens* has been interfering with the genetics of plants and animals. The tiny, stentorian teosinte, a grass produced by natural selection, has turned into the massive, calorific corncob thanks to the artificial selection performed by generations of farmers. The Chihuahua, the smallest lapdog, derives from the artificial selection performed by generations of breeders starting from a quite different product of natural selection: the wolf.

In more recent decades, *Homo sapiens* has been interfering more profoundly with the genetics of plants and animals. In 1953, it was discovered that all life is reproduced using the same nitrogenous base alphabet ATCG (adenine and thymine, cytosine and guanine), arranged in a sequence of pairs in a molecule of deoxyribonucleic acid, better known as DNA. In 1994, the first genetically modified, longer-lasting tomato hit the supermarket shelves in the US. In 1996, the first mammal, Dolly the sheep, was cloned. In 2001, scientists first decoded the human genome, made up of 3,088,286,401 pairs of nitrogenous bases. This enterprise may have required over three billion dollars in investment but, by 2018, it had become so efficient that it could be achieved for close to the fabled threshold of 1,000 dollars.

What is likely to happen in the next century or two is truly unfathomable. It ranges from the dystopian scenarios already described in literature to a much rosier future imagined by **transhumanism**, an international movement that champions methods and technologies aimed at increasing lifespan indefinitely and expanding brain potential until the human species earns the name 'post-human'. There's no need to remind ourselves that, in both cases, there are monumental ethical obstacles to overcome. These are some of the most formidable challenges awaiting the still-human species in the future. Because there's a rule of sorts underlying the characteristic *sapiens* curiosity: if something can be done then somebody will want to do it.

Just think of **optogenetics**, one of the most extraordinary neurotechnologies to have emerged in recent years. It is incredible that someone even took it upon themselves to deploy the suggestion made by Francis Crick: in order to control single neurons, the co-discoverer of DNA wrote in 1999, 'light is the ideal signal'. With magnetics and electricity, science had already found a way of influencing whole areas of the brain, but not individual neurons – optogenetics has managed that The process starts off by isolating from algae and bacteria the genes that express various kinds of opsins [p102], light-sensitive proteins. These genes are then inserted into the DNA of lab mice, so that different opsins may respond to different neurons. The brains of these mice are then connected to optic fibres that carry light at various frequencies. *Et voilà*, by simply modulating the frequency of the light (blue, red or yellow) individual neurons can be inhibited or stimulated, manipulating the behaviour of the mice as though they were remotely controlled. Optogenetics, which helps us to understand the function of individual neurons, is so revolutionary and promising that, although it has only recently emerged, it is already being used by hundreds of laboratories around the world.

In the meantime, another much more powerful and revolutionary technology has appeared, which, according to some, is destined to change not only scientific research, but the world as we know it. It's called **CRISPR-cas9**. In a nutshell, it allows us to cut and paste DNA so easily, quickly and inexpensively that even ten years ago nobody would have believed it.

Bacteria and viruses have been waging their daily battle for survival for much, much longer than lions and gazelles. Consequently, some bacteria have developed a complex system for 'stealing' parts of the DNA from the viruses that attack them, in order to recognise the viruses and defend themselves against them next time the situation arises. Employing exactly the same system, scientists use enzymes that bind themselves to the DNA, cutting the DNA at a specific point in the chromosome, where a gene is replaced with another, and then sewing the cut back up.

The great thing is that it works a treat.

The bad thing is that this technology is so easy and so inexpensive that it can be used for ethically questionable and undoubtedly dangerous purposes. In 2015, a team of Chinese scientists from the Sun Yatsen University performed CRISPR-cas9 experiments on human embryos only to abandon them afterwards. The ethical line that was once not to be crossed has already been crossed. The following year, James Clapper, the US Director of National Intelligence, included CRISPR-cas9 in the list of six major planetary risks, along with North Korea and Russian missiles. Why? Because the cutting and pasting of genes can also be used to manufacture devastating biological weapons.

Venturing a little further into the future, the great promise of genetic modification is the eradication of genetic disorders. For the time being, though, the manipulation of the human genome is considered unacceptable. Moreover, the Chinese team who used CRISPR-cas9 on embryos came across far more undesirable effects than anticipated. But will it be the same in 20 or 40 years' time? Currently, the functions and interrelations of genes are still largely unknown, but when they become clearer, will society agree to use them for the sake of curing cystic fibrosis or Huntington's Disease? Will it *only* use them to this end? Or will it make the even less ethical decision to continue to allow those who are affected by these conditions to suffer?

In our survey, we come finally to the aberrations of so-called genetic cosmetics: parents who choose the exact eye colour, say, of their future baby from a catalogue. If something can be done, then somebody will do it. The other possibility, not exactly cosmetic, is that we might eventually manage to get our hands on intelligence genes. All it takes is for somebody to discover how to do it, and private clinics for enhancing the cognitive abilities of people's children will spring up like mushrooms. As ever, it will be the market that decides. Will wealthier parents resist the temptation to determine in advance that their daughter will be an ace at mathematics and a genius at the piano? Will genetic clinics go bust through a lack of clients? You can imagine your own answer to that.

As we have said, for thousands of years, *Homo sapiens* has

been interfering with the genetics of plants and animals, but this is only the beginning. At some point, genetically modifying intelligence will be an overpowering temptation.

Nowadays, we can think what we like about it. It will be a choice presented to future generations. From the point of view of the evolutionary history of the human race, this would be another historic upgrade (4.3.9) of the system version [p10].

12.3 ARTIFICIAL INTELLIGENCE

As we know, exams never end. But don't fret about the Turing Test too much. If you ever had to take it, you would pass it easily even without preparing for it. And you would beat all the calculators, including the IBM Summit, which, at the end of 2018, is the most powerful supercomputer in the world (122 million billion operations per second).

This test, devised in 1950 by Alan Turing, helps evaluate the intelligence of a computer and so far no computer has ever passed it. The idea of the British scientist, whose amazing life and tragic end are told in the film *The Imitation Game*, is very simple. In order to be considered intelligent and 'thinking', a computer must make a human being believe that it's human.

Although humans have long fantasised about it, the term 'artificial intelligence' has a date and place of birth: summer 1956, Dartmouth College, New Hampshire. A small handful of computer scientists got together for six weeks to draft the theoretical bases for the thinking machines of the future, and christened the topic **artificial intelligence**, or AI. Ten years later, their research was very generously financed by the US government, and the Department of Defence in particular, encouraging the scientists into a somewhat excessive leap of faith. 'Within a generation… the problem of creating artificial intelligence will substantially be solved,' said Marvin Minsky, one of the Dartmouth conference delegates.

It didn't work out that way. For decades, the fate of AI has fluctuated. The first public success occurred only in 1996, when IBM's Deep Blue computer beat the chess world champion Garry Kasparov. In 2011, another IBM computer, Watson, beat two legendary champions of *Jeopardy!*, the linguistically challenging American quiz. Yet neither of the two machines, each built around a furious computational capacity and a gigantic database of linked information, has really passed the Turing Test.

Nonetheless, all of a sudden AI has started to truly become part of people's daily lives. Machines are learning to learn – it's called **machine learning**, automated learning. While Watson followed an intricate set of instructions, but was unable to alter them, AlphaGo can do so. AlphaGo is a software program written by DeepMind, a London company set up by Demis Hassabis and acquired by Google in 2014, which beat the world champion of Go, considered to be the most complex game in the world (the possible combinations are 2×10^{170}: much, much more than the number of atoms in the universe). After instructions and data were inserted by a human hand, AlphaGo was able to learn, thanks to a **deep neural network**, a branch of machine learning based on a series of algorithms that perform operations on many levels, simulating the hierarchical layers of the cerebral cortex. What's most amazing is that AlphaGo self-built its own abilities – the way humans do – by playing 30 million games with itself and learning from its own errors.

Machine learning and neural networks also underlie Siri, Cortana, Alexa, Ok Google and co., the vocally commanded personal assistants installed on all smartphones, and now also on special automated household devices. The noteworthy fact is that they learn from users' requests, and consequently improve over time.

AI now has a firm place in the automotive industry, thanks to the Israeli MobilEye (founded by scientist Amnon Shashua and acquired by Intel for 15.3 billion dollars), which was the first to develop a system of intelligent vision to enable automatic safety. Driverless cars, on which all the automobile giants, as well as Tesla, Google, Apple, Uber and many others are working, could become a reality within just a few years. The plan is to literally delegate the steering wheel to AI.

AI has already caught on in factories, where new kinds of robots collaborate with human operators and learn the most varied tasks from them. There are already algorithms based on machine learning that are able to write some legal documents without the need of a lawyer, a sports or finance article without the need of a journalist, or musical arrangements without the need of a composer. 'The professions less likely to be replaced by machines,' says Tomaso Poggio, professor at the Massachusetts Institute of Technology, who counted Demis Hassabis and Amnon Shashua among his postdoc students, 'will be the simpler but more creative ones (plumber, handyman) and the more complex ones (scientist, programmer). All the others will be largely replaced.' Even though politicians currently attribute national unemployment to the problems of the global market rather than to growing automisation, true statesmen (the ones concerned with the future as well as the present) must gradually prepare for the impact of AI on society. It has already begun.

Deep learning has become possible thanks to the convergence of three factors: the constant increase in microprocessors' computational power; the development of new techniques and more sophisticated new algorithms; and the availability of large databases to train the muscles of AI, as in the case of AlphaGo. Of these three factors, only the first advance could be finite. Moore's Law ('processing power doubles every two years') is about to exceed the physical limitations in silicon chips, to the point that in the last few years AI has been relying on GPUs, graphics microprocessors, which perform multiple computations simultaneously in parallel and are more efficient. For Moore's Law to continue, we need something else. For example, **neuromorphic chips** that simulate the brain.

It's an old idea but probably nearing practical realisation. Traditional processors perform calculations at the rhythm of a clock, a metronome that beats their time. In the brain-simulating processor, however, besides communicating in parallel without time constrictions, every artificial 'neuron' is able to receive information and decide whether to transmit it to the next neuron. Just like real neurons. In addition to this, a neuromorphic chip uses very little energy, exactly like the brain: a prototype, again built by IBM, contains five times as many transistors as an Intel processor, but uses 70 milliwatts, 2,000 times less energy.

There are no obstacles to the growth of the other two drivers behind the machine learning boom – increasingly sophisticated algorithms and large databases. Even so, one thing must be clear: we're talking of a technology that is still primitive. Neural networks manage to dig out complex statistical connections in real forests of data, but not much more than that.

AI promises to become an unstoppable wave of technological progress. As with genetic modification, there are those who idolise it and those who see the devil in it. A **technological singularity** is defined as a radical change in human society – the exact moment when an artificial super-intelligence will trigger the beginning of a technological growth never before known in history. Ray Kurzweil, head of technology at Google and author of the book *How to Create a Mind*, is one of its most passionate supporters. But famous entrepreneurs such as Bill Gates and Elon Musk, together with the late Stephen Hawking and other influential scientists, share the view of the Future of Life Institute of Boston, which is busy warning the opposite: AI is an 'existential risk'. It could put the human race in danger.

It is not up to a simple manual to express philosophical opinions, which call for much weightier and more authoritative books and publications. Even so, we will take the liberty of raising a question. Even before this technology is available, shouldn't what can be done with it be a cause for concern? Bionic and robotic soldiers are not things of fantasy: the military laboratories of superpowers have been working on them for years. The possibility of using CRISPR-cas9 to devise terrible biological weapons did not come out of science fiction books, but out of the mouth of the head of US intelligence.

Finally, the possibility that madmen, criminals or terrorists may trigger a large-scale cyber war – since communications, airlines, aqueducts and hospitals are connected to the Web – is sadly more plausible than we think. Add to that nuclear weapons and climate change, and the dangers of AI are in good company.

Over the past decades, thanks to digital communication, human brains have become connected like never before in history, giving rise to a kind of planetary intelligence, which, in a supra-national scientific environment, is yielding enormous results. And yet, in the growing complexity of a world that is home to eight billion brains, could we, perhaps, do with a higher dose of intelligence?

One thing is certain. Artificial intelligence – with precautions, we hope – will be pursued regardless of protests against it, because of a natural, endlessly curious intelligence that evolved from the primitive brains of half an billion years ago.

It will be system version 5.0.

Appendix

GUARANTEE
This product is not covered by any form of guarantee or extended, limited or partial warranty.

 WARNING
Prior to using this product, you take on all the risks and responsibilities related to its usage.

Except where explicitly forbidden by current legislation, you are free to transport and use the product anywhere you like, as long as you respect the thermal (36-37 degrees Celsius), altitude (5,000 metres) and pressure limits. Beyond its operational limitations, there are no significant effects on the non-guarantee, but we must warn you that those on the central nervous system could be fatal.

In the event that you think the product is malfunctioning, please contact directly the specialised help centres listed on https://www.nhs.uk.

Insurance policies are available on the market, but we recommend you pay special attention to the clauses in the contract. You will not be offered a replacement, just a partial financial refund, payable only to your next of kin or chosen charity.

TROUBLESHOOTING

The brain won't turn on	Check again. If you're reading these words it means it's on. Then make yourself a coffee.
Won't start up	The reset button is disabled in this version. Try performing a whole cycle in stand-by mode [p87].
Won't switch off	This system version is *always on* and must never be switched off. Instead, please consult the instructions for stand-by mode [p87]. Warning: the system switches itself off only at the end of the product's life [p211] and, at present, cannot be reactivated [p233].
Won't go into stand-by	Follow the instructions on pages 87 and 93. If, after trying for 48 hours, it does not go into stand-by mode, urgently seek medical assistance.
The image is blurred	If you usually use corrective lenses (optional), check that they are fitted correctly. Otherwise seek medical assistance.
The screen is completely black	Try checking for the presence of photons in the room. If it's still totally black, check that there isn't an electrical black-out. If not, immediately seek medical assistance.
The audio can't be heard very well	If you usually use auditory silencers (optional), check that they are uninstalled. If not, seek medical assistance.
Short-term memory is too short	Read the instructions on pages 64 and 221.
Is unable to learn	That is effectively impossible. The brain is a *learning machine*, starting as soon as it leaves the factory. However, if you have never performed the unblocking procedure, known as *growth mindset*, go immediately to page 72.
Motivation seems jammed	Follow the recommendations [p154] and repeat the cycle several times and unblocking is 100% guaranteed. In the unlikely event that it is not, check the terms and conditions of your guarantee [p248].

Crashes easily	If you're referring to overexcitement, a fast heartbeat and possibly sweating at the peripheral articular terminals, we recommend you keep stress under control [p196]. If, on the other hand, you're referring to episodes of anger (the proverbial 'being pissed off') please see page174. In both cases, breathe deeply, slowly and rhythmically for one minute.
Menu cannot be found	Your brain is completely automatic so does not require a menu – for instance, you don't need to select the 'run' function from a drop-down in order to catch to the departing train. The high-speed connection between the cerebral cortex and the lower limbs ensures an automatic transfer of the mental commands in under 100 milliseconds. If making a selection from a main menu were required, you would certainly miss your train.
The 'thought' function seems to have slowed down	Always make sure that you maintain the basic efficiency of the system with sufficient stand-by periods, regular insertion of water and healthy foods and steady physical exercise. If you have any doubts about this advice, please see page 93.

LEGAL DISCLAIMER

The title of this book, *The Brain: A User's Manual*, is intended strictly as a narrative device or a *divertissement*, and not to be taken literally. The aim of this little book is to inform the average *sapiens* brain user of the main characteristics of the cerebral machine in his or her possession, through facts, ideas and reflections that could be useful throughout his or her daily cerebral experience.

In no way is it our purpose to tell anyone how to treat disorders or to offer advice or to supply a self-help guidelines.

We are also obliged to inform you that this manual was written by an ordinary journalist, genuinely passionate about science but a veteran of studies in the humanities. This information does not entitle you to terminate this agreement.

The content of this book is entirely the responsibility of the author. It was he who dismissed the idea of increasing the girth of this manual with pages and pages of notes about sources used: these sources are the books mentioned in the text and the select bibliography [p257], as well as many online peer-reviewed scientific articles (published in the likes of *Science* and *Nature*) and popular science from Wikipedia, YouTube, Ted, Coursera and Khan Academy. In the attempt to simplify in just a few pages the most complex thing there is, the author has tried to choose the most interesting information and, out of the myriad scientific controversies, those positions that have the widest consensus among experts and, in very few cases, those he liked most.

The final product may not be immune from cognitive biases [p189]. On this subject, let it be clear that the author's responsibility is restricted to exclusively moral boundaries and does not carry practical responsibility for an improper use of the product [p248]. This is governed by Section 1 of the Civil Court of Timbuktu, Mali.

Do not keep this manual out of the reach of children.

Afterword

by Tomaso Poggio
Director, Center for Brains, Minds and Machines, MIT

The origin of the universe. The structure of matter. The mystery of life. The evolution of intelligence. These are four of the greatest questions that are facing modern science and that will keep scientists busy for decades and maybe centuries to come. I believe intelligence is the challenge of this century, just as physics was during the first half of the 1900s and genetic biology in the second.

However difficult it might be, understanding and replicating intelligence is no doubt a crucial challenge. Progress in solving the problem of intelligence will allow us to increase both our intelligence and that of computers, helping us to solve science's other great problems more easily.

Just a hundred years ago we believed that the Milky Way was the universe, until Edwin Hubble showed us that it's only one of about 200 billion galaxies. Only 70 years ago, we did not know how the marvel of hereditary transmission worked, until Francis Crick and James Watson revealed the hidden secret of every cell. The recent evolution of science has almost doubled our average lifespan and multiplied our knowledge. In other words, it has added more blocks to the evolution of the human species.

It is clear that natural selection has shaped modern man

using the same instruments that produced ferns and baobabs, insects and elephants: the genes. But genes alone cannot explain the evolution of human intelligence. There are also ideas, which biologist Richard Dawkins calls memes. Memes, just like genes, can either compete or collaborate, they can preserve themselves and they can mutate. These memes spread like a virus: they replicate, evolve and select themselves.

The technologies developed by humans – starting with fire and the wheel – have become an integral part of their very evolution, which is by now inextricably linked to cultural and technological evolution. From this point of view, we could say that the human race has gradually been equipped with a kind of *superbrain*, a global intelligence that goes beyond the individual one. Just to give an example, there is no person in the world capable of understanding (and manufacturing) all the microchips, all the feeding and communication systems, as well as all the software that makes up a modern smartphone millions of times more powerful than the computer used by NASA for the first Apollo missions.

We are at the historical moment when the role of artificial intelligence, and *machine learning* in particular, has officially begun. The fusion of neuroscience with computer science is destined to imbue machines with increasing doses of intelligence which could turn out to be critical for public health, education, security, and, in general, for the prosperity of a world called upon to make difficult choices. I am certain that, in the future, artificial intelligence will make a useful contribution to the collective decisions that will be needed to tackle the dilemmas of a planet that has been transformed – not always for the better – by the human race.

Some people might think that this 'cosmic' and futuristic view of the evolution of intelligence has little to do with a manual for the brain user. I think the contrary. It is the knowledge accumulated by generations and generations of brains that has designed the world as we know it. And yet, as Marco Magrini observes, the individual user of the neural machine is usually not well informed about the mechanisms that make it work. Its extraordinary mechanisms, sometimes unthinkable and counterintuitive, are increasingly well understood by neuroscience, but remain obscure to the majority of people.

In the 21st century, this gap must be closed. Schools currently design the content of education without teaching what this intricate biological learning machine is or how it works. Knowing the electrochemical processes behind not only emotions and feelings, but also motivation and creativity, takes nothing away from the pleasures (or displeasures) of life: and it could be really crucial to a more self-aware existence. I personally wish that an increasing number of people, starting with politicians, knew what neuroscience knows today, and were prepared to open their minds to what we will likely know tomorrow.

Just as Copernicus, Galileo and Hubble changed the concept of our place in the universe, an extraordinary harvest of neuroscientific discoveries is now shining a light on the most complex and amazing thing in the universe: the human brain. This book, which has the additional good fortune to be entertaining, summarises it beautifully.

Select Bibliography

Brynjolfsson, Erik and McAfee, Andrew, *The Second Machine Age – Work, Progress, and Prosperity in a Time of Brilliant Technologies*

Burnett, Dean, *The Idiot Brain: A Neuroscientist Explains What Your Head is Really Up To*

Carr, Nicholas, *The Shallows: How the Internet is Changing the Way We Think, Read and Remember*

Crick, Francis, *The Astonishing Hypothesis: The Scientific Search for the Soul*

Damasio, Antonio, *Descartes' Error: Emotion, Reason, and the Human Brain*

Dawkins, Richard, *The Greatest Show on Earth: The Evidence for Evolution*

Dennett, Daniel, *Freedom Evolves*

Doidge, Norman, *The Brain That Changes Itself: Stories of Personal Triumph from the Frontiers of Brain Science*

Duhigg, Charles, *The Power of Habit: Why We Do What We Do, and How to Change*

Dweck, Carol, *Mindset: Changing the Way You Think to Fulfil Your Potential*

Foer, Joshua, *Moonwalking with Einstein*

Kurzweil, Ray, *The Age of Spiritual Machines: When Computers Exceed Human Intelligence*

Kurzweil, Ray, *How to Create a Mind: The Secrets of Human Thought Revealed*

Marcus, Gary. *Kluge: The Haphazard Construction of the Human Mind*

Markoff, John, *Machines of Loving Grace, The Quest for Common Ground between Humans and Robots*

Mlodinow, Leonard, *Subliminal: The New Unconscious and What It Teaches Us*

Newport, Cal, *Deep Work: Rules for Focused Success in a Distracted World*

O'Shea, Michael, *The Brain: a Very Short Introduction*

Oakley, Barbara, *A Mind for Numbers: How to Excel at Math and Science*

Punset, Eduardo, *El alma está en el cerebro*

Ridley, Matt, *Nature via Nurture: Genes, Experience and What Makes Us Human*

Sacks, Oliver, *Musicophilia: Tales of Music and the Brain*

Sapolsky, Robert M., *The Trouble with Testosterone: And Other Essays on the Biology of the Human Predicament*

Shermer, Michael, *The Believing Brain*

Tononi, Giulio, *Phi: A Voyage from the Brain to the Soul*

Wright, Robert, *Nonzero: The Logic of Human Destiny*

Acknowledgements

In 2013, when, after 24 years, my brain decided to put an end to my experience with the Italian business daily *Il Sole 24 Ore*, it felt the desire to explore the subject of artificial intelligence in greater depth. So I turned to Professor Tomaso Poggio for help. He is one of the fathers of computational neuroscience and I had met him just a few months earlier during an interview. He was kind enough to let me spend three months at his laboratory at the Massachusetts Institute of Technology and became my friend.

During that time, immersed in the mechanisms of intelligence as never before, the aforementioned brain produced a thought totally automatically: 'We have a user's manual for everything, from the fridge to the electric toothbrush, but not for the most important machine we own.' That's how the idea for this book – disarmingly silly as it was – came about.

This is why my first thank you goes to Tommy, as everybody calls him.

The second goes to his wife, Barbara Venturini-Guerrini, who, in addition to welcoming me like a sister, read this manual as it was being written (she is a neuropsychologist) and gave me advice as well as the right dose of dopamine for my motivation.

The next thank you goes to four other friends who formed my trusted reading group: Maria Briccoli Bati (neurophysiologist), Valeria Marchionne (editor), Miriam Verrini (journalist) and Pietro Tonolo (musician).

Thanks to Todd Parrish and Daniele Procissi, professors at Northwestern University in Chicago, for their explanations of fMRI technology (and for taking me on a tour of it). Thanks to Professor Andrea Camperio Ciani of the University of Padua for his advice on brain models. Thanks to 'my' editor Veronica Pellegrini and to Laura Venturi.

Thanks for encouragement and discussions also go to (in random order) Anna and Alberto Miragliotta, Anna and Piergiorgio Pelassa, Annalisa and Andrea Malan, Valentina and Aldo Gangemi, Maurizio Bugli, Alex Jacopozzi, Monica Mani, Isabella Bufacchi, Rossella Ballabio, Cesare Peruzzi, Graeme Gourlay, Annamaria Ferrari, Eleonora Gardini, Marco Pratellesi, Patrizia Guarnieri, Luca Magrini, Giuditta Gemelli, Pierre de Gasquet, Celio Gremigni, Alessandro Bronzi, Piero Borri, Massimo Ercolanelli, Francesco Maccianti and many more, including my dear classmates and my children, Jacopo and Carolina, to whom this book is dedicated.

A special recollection for Marco Lamioni, refined musician and kind man, who had so much fun listening to me talk about this book on the brain, even though that was precisely where disease had struck him.

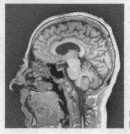

Brain scan courtesy of Northwestern University,
Neuroimaging Reserch Lab, Chicago

About the Author

MARCO MAGRINI is a science and technology journalist, born in Florence, Italy. He claims to be a descendant of Giovanni Villani, an eminent chronicler of medieval times. After working 24 years with the Italian daily *il Sole 24 Ore* as a senior correspondent, he now contributes to *La Stampa* newspaper and writes the column 'Climatewatch' for *Geographical*, the Royal Geographical Society's magazine.